P9-CRF-659

The Happy Atheist

The Happy Atheist

PZ MYERS

PANTHEON BOOKS · NEW YORK

Library of Congress Cataloging-in-Publication Data
Myers, PZ (Paul Zachary)
The happy atheist / PZ Myers.
pages cm
ISBN 978-0-307-37934-4
1. Atheism. 2. Religion—Controversial literature. I. Title.
BL2747.3.M94 2013 211'.8—dc23 2013000202

www.pantheonbooks.com

Jacket image: A variant of the Darwin fish
by Evolution Design, Inc.
Jacket design by Brian Barth
Book design by Maggie Hinders

Printed in the United States of America

First Edition
2 4 6 8 9 7 5 3 1

To Mary, Alaric, Connlann, and Skatje.
Not that they'd agree with everything I say,
but I'm deeply appreciative of a life lived among
argumentative and irreverent freethinkers.

When I became convinced that the universe is natural, that all the ghosts and gods are myths, there entered into my brain, into my soul, into every drop of my blood the sense, the feeling, the joy of freedom. The walls of my prison crumbled and fell. The dungeon was flooded with light and all the bolts and bars and manacles became dust.

—Robert G. Ingersoll, "Why I Am an Agnostic"

Contents

Morning in the Midwest · 3

About the Author · 10

The Joke · 14

The Great Desecration · 18

Get a Job · 28

It's So Easy to Be Outraged! · 31

I Am Not a Spoiled Child Having a Temper Tantrum · 34

The Proper Fate for a Holy Book · 39

Ask but Don't Tell · 42

Dirty Words · 47

The Top Ten Reasons Religion Is Like Pornography · 51

The Purpose-Free Life · 52

Happy Easter! · 58

Afterlife? What Afterlife? · 62

Soulless! · 65

What Dreadful Price Must We Pay to Be Atheists? · 69

Imagine No Heaven · 74

Daughters of Eve · 82

Prometheus's Sin · 91

So Alone · 95

One Nation Free of Gods · 100

An Embryo Is Not a Person · 104

The Courtier's Reply · 109

The Big Pink Guy in the Sky · 112

The Karen Armstrong Diet · 121

God's Little Crisis of Confidence · 126

Laughter as a Strategy for Diminishing Religion · 131

We're Happier out of a Straitjacket than in One · 137

Marketing Godless Science · 141

"Science Is What We Do to Keep from Lying to Ourselves" · 146

Our Brains Are Full of Contradictions · 149

Ken Miller, Poster Child for Compatibility · 151

Religion Fails as a Source of Knowledge · 157

Science as a Lever to Move the World · 159

The Active Hand · 161

The Proper Reverence Due Those Who Have Gone Before · 167

Niobrara · 175

We Stand Awed at the Heights Our People Have Achieved · 180

Acknowledgments · 189

The Happy Atheist

Morning in the Midwest

On any fine morning in rural Minnesota, I can step outside the door of my home and look a few blocks to the southwest and see the Church of the Assumption of the Blessed Virgin Mary. A few blocks to the west, just out of sight behind nearby houses, lies the First Lutheran Church. About four blocks to the east is the Federated Church, the "liberal" church in town. Even closer stands the Lutheran Campus Ministry, which serves the university at which I work, and the Newman Center, its Catholic counterpart. Since this is Minnesota, I could make fairly fine-grained sectarian choices within Lutheranism: the First Lutheran Church belongs to the Evangelical Lutheran Church of America, but I could attend Faith Lutheran Church, which is another member of ELCA. Or if I wanted something a bit more conservative, I could attend St. Paul's Evangelical Lutheran Church, which belongs to the Wisconsin Synod, or Bethel Evangelical Lutheran Church, of the Church of the Lutheran Confession. There's also the Zion Lutheran Church nearby, which belongs to the Missouri Synod.

If I were really broad-minded, I might choose the First Baptist Church, the Kingdom Hall of Jehovah's Witnesses, the Apostolic Christian Church, or the Morris Community Church, which meets in the local high school. I count fifteen churches within walking distance of my house; there are no synagogues or mosques, probably because the believers they would prey upon are too thinly populated here to be profitable.

Often when I step outside, I am treated to hymns played at a stunning volume from the loudspeaker in Calvary Cemetery, just two blocks to the north. During the day, they sound at least every hour—when the old fellow who donated the loudespeaker to the cemetery is feeling particularly pugnacious, they'll be programmed to play every quarter hour. The coffee shop downtown is owned by a coalition of churches, which normally makes no difference to me—the coffee is secular, at least—but I've learned not to visit it on Tuesday mornings, when the Men's Bible Study group meets there. It is simply too stressful to try to hold my tongue when they discuss the historicity of Noah's Ark, or the imminence of the Rapture and the End Times, while I sip my coffee. I've succeeded in keeping my thoughts to myself so far, but I wouldn't want to risk an eruption. At some moments, like the time they were discussing the proper role of Christian women (that is, to be subservient and quiet), I have been tempted to perform a few secular baptisms with that good secular coffee.

This town has a total population of roughly five thousand people. As I said, we have fifteen churches, but only one bookstore, which doubles as the video rental store, the kids' card game store, and the local cell phone retailer. We're drowning in piety with a host of preachers, each one of whom offers a different version of the eternal and irrefutable truth, each one of whom takes advantage of the tax-exempt privilege of religion to build a little temple to his or her peculiar dogma.

There is nothing unusual about my town. This is perfectly ordinary, rural midwestern America, like thousands of other small towns all across the country. We're just immersed in religion, like every other God-soaked spot in lightly populated, Republican-leaning, Real Live Genuine USA.

My fellow townsfolk are good people, like most human beings, who are mostly concerned with getting along, doing well for their families, and seeing their community thrive as a safe and stable place. I don't accept the common atheist line that religion makes people do evil acts, like fly airplanes into buildings and start holy wars; it can and has done so, of course, but those are the pathological extremes, and it isn't right to judge an idea by the excesses of the maniacs who turn it into a cause for violence. Mainly what religion does is make people believe ludicrously silly things, substitute dogma for reason and thought, and sink into self-destructive obsession as they fret more over their reward in the next life than their accomplishments in this one.

There is also a mix of levels of fervor in town. Certainly there are zealots who are convinced that every fornicator is going to burn in hell for eternity, but I suspect that the fornicators outnumber the fervent. For every true believer, there may well be a dozen who go to church simply because they've always gone to church, because they're in charge of bringing the coffee cake to the social afterward, because it's been dunned into their heads that good people express their worth publicly in regular attendance, because they've been told it provides good moral guidance for their children, or because their mother-in-law would make family get-togethers a living hell if there was even a hint that their faith had lapsed.

But some of these sincere believers show hints of desperation with which I can sympathize. The old man obsessively poring over the pages of his Bible, his finger moving word by word over Revelation, quietly mumbling—he is afraid, his mortality is hanging over him, and he wants to make sure he's right with the Lord, whom he's sure he's going to face soon. The woman wracked with guilt over her wild teenage years, her abortion, her flirtations with drugs and sex—now her kids have reached that same age, and she can't bear the thought that they might carry out the same experiments she did, and she hopes a little hellfire and Christian discipline will keep them in line. And so many people have secular concerns: Will the economy improve? Will America triumph? Will their children prosper? They've been told so often that the fate of the nation lies in courting the goodwill of God that they seek answers to real-world problems in the reassuring fantasies and apocalyptic terrors of faith.

I'm an atheist swimming in a sea of superstition, surrounded by well-meaning, good people with whom I share a culture and similar concerns, and there's only one thing I can do.

I have to laugh.

Living in America at the beginning of the twenty-first century is like attending the circus when the clowns are performing—it's low comedy, full of pratfalls and pies in the face and silly costumes. It's also hilarious.

The people all around me seriously believe most fervently in a god who briefly became human and was tortured to death for his trouble—and that this is the greatest story ever told. The omnipotent Lord of the Cosmos is a man-like entity who, in addition to keeping the planets in their courses, deciding the fates of nations, and spawning hurricanes and earthquakes, frets endlessly about the sex lives of his chosen people, insisting that the males chop off a bit of membrane at the tip of their penises and that the females preserve a scrap of membrane in their vaginal canals as long as possible. Lately he has become a devotee of football, and players and spectators beg for his divine favor in helping to get a ball from one side of the field to the other. His followers are offended at the thought that they have distant relatives who are monkeys, but they feel ennobled by the myth that they were made from dirt by an invisible hairy anthropoid ape in the sky. He is a soap-opera god who instills in his people a disproportionate sense of propriety that isn't fitting for the high-minded imagined philosophy of their prophet, Jesus, let alone for a two-year-old child. The faithful are fond of claiming that he is a god of peace and love, but most often he seems to be a god of nosy busybodies, prying prudes, and outraged promoters of arbitrary pettiness. Jehovah is the god of sanctimonious killjoys. How else to explain people who think the greatest threat to their happiness right now is the existence of people of the same sex who fall in love with each other? And at the same time, we have priests raping children, and their defenders are most irate at the people who ratted out the pedophiles in dog collars.

I try to laugh, but I also feel the human suffering caused by these follies. But then, as Shakespeare knew, the best comedy has always been born out of the truth of pain and has a leavening of tragedy mixed in with it.

I confess, sometimes the scampering of the clowns gets wearing: the humor is too crude, too over-the-top, and sometimes you want something a little more subtle and sophisticated. And when that happens, I can always reliably turn to the po-faced commentators, the apologists in the bleachers who watch absurd religious rituals performed before them and, instead of laughing, offer excuses and interpretation. Religion is a serious business, they say, as the sick line up for miracle cures at Lourdes, and transubstantiationists argue that wafers of bread really do turn into the substance of a god in the mouths of their flock, and rabble-rousing politicians assure their followers that they didn't come from no monkey and that, yes, God created the earth at a date we've since determined occurred nine thousand years after smarter, more practical human beings domesticated the dog. This is another kind of humor, the spectacle of seeing smart, well-educated people suspend critical thought in order to invent elaborate rationalizations for indefensible absurdity.

It's weird enough to watch people pantomime trying to escape an invisible box, but it adds a whole new level of meta-weirdness to see other people straining to argue that the box is actually there.

So I laugh, and I admit, the laughing often seems mean-spirited. Look at the people wasting their lives praying, of all things, or wracked with guilt because Iron Age goat herders regarded their harmless desires as sinful! Aren't they stupid? No, they aren't, and I know they aren't; I don't regard myself as smarter than the townsfolk around me. What I see are ordinary people afflicted with folly, who could be happier and more productive if they were free.

People aren't religious because they're stupid; rather, religion is a parasite of the mind that makes people do stupid things and think stupid thoughts and, worst of all, believe that freeing themselves of superstition would make them less human and less able to cope with reality, when the exact opposite is true. I laugh because they do goofy, foolish things in the name of their god, but also because I hope that a little raucous hilarity will wake a few of them from their surreal lives and show them how wonderful they and the real world are, if they only lived in it.

That's what this book is about: the folly of faith, the absurdity of apologetics, and, I hope, the joys of reality, as seen through the focusing lenses of humor as well as the more traditional ones of science and reason.

About the Author

I know this usually comes at the end of a book, but I thought it was important for you to know who you are dealing with. You may not believe me, but I swear it's true: I'm not of this earth. I fled here years ago because my home planet was driving me crazy. Let me explain.

My home world is very much like this one. It's populated by billions of bipedal primates, who are just like people here: sometimes foolish, sometimes wise, sometimes hateful, sometimes generous. They are grouped into cities and nations, and sometimes they have wars, and sometimes they cooperate. You really would have a hard time telling our two planets apart, except for one thing.

The hats.

My people are obsessed with hats. Almost everyone wears them, and a lot of their identity is wrapped up in their particular style. Some people always wear cowboy hats, for instance, and others wear bowlers, and each thinks the other is exceedingly funny-looking and would never consider switching. They have elaborate ceremonies for their children in which they confer the hats, and kids often go to special schools once a week where they learn about the history and significance of their hats. From birth to death, everyone has the importance of hats drilled into them.

The particular type of hat one wears is critical. Individuals only rarely change hat styles, and when they do, those who wear the abandoned style consider it grounds for sorrow, while those wearing the newly adopted style rejoice. Sometimes people invent new kinds of hats, which are typically regarded as bizarre when one person is wearing it, but once a sufficient number switch to the new style, they are respected automatically. The streets of our more cosmopolitan cities are filled with strange and comical hats bobbing along, but no one laughs. Laughing at a hat is considered a heinous crime.

It sounds very silly, I know. A minority on my planet also find it pointless, myself among them, and don't bother wearing a hat at all. Hatlessness is tolerated in the more civilized nations, but in some places wearing no hat, or a strange hat, can get you killed. Also, many people in my country bother to wear their hat only once a week. The rest of the time they keep them on ornate hat stands in their home and attach much significance to their presence.

Now why should a mere excess of fashion compel someone to travel many light-years to escape? There is something more. Most people adhere to a remarkably absurd notion: that an important portion of their minds actually resides in their hats. They believe the locus of their ethical sense is somehow intertwined in their hats' fabric. This belief leads to strange customs: witnesses in trials are required to wear their hats to give testimony; soldiers are thought to be cowards without their hats; politicians vie to see who can wear the most ostentatious hat; sex is considered a filthy practice because people will take off their hats to do it. Hallowed by tradition but devoid of scientific evidence, the custom persists.

Hatters, milliners, and haberdashers are highly regarded professionals, and every town has numerous hat shops. The person who crafts miters doesn't also make berets, or vice versa, but both still prosper because not only do the majority of people sink a significant proportion of their income into the purchase and care of their hats, but the occupation is considered too dignified to be taxed. Huge sums of money are poured into hatteries, and the people consider wearing the proper hat to be noble and right. The president of my country listens very closely to his council of hatters, and no television punditry is complete without a haberdasher to use his vast hat-based wisdom to pontificate on domestic and foreign policy. They are all talking out of their hats, which is considered a very good thing.

I couldn't help noticing, though, that the very idea that ethical thought is localized to a hat is ridiculous, and that hatless people can be just as good and kind and wise as those with the most ornate hat (and that hatless people can also be wretched and cruel, of course, as can the hatted). Our president has a rhinestone-covered twenty-gallon cowboy hat with an air horn and flashing strobe, and he seems far less virtuous than my neighbor, with her simple and unostentatious cap. Hats obviously have nothing to do with morality, except perhaps in an inverse sense: those who spend the most effort polishing the gewgaws and flash on their hats usually put the least effort into honing their minds.

I could see the writing on the wall. Being hatless myself meant my chances for promotion were limited, but even more worrisome, the height of one's hat was becoming the sole measure of nobility of purpose, and the genuine leaders were being replaced by loud poseurs who knew how to stretch a crown and use a BeDazzler. When the People of the Easter Bonnet started encouraging war with the Chador Wearers, citing deep philosophical differences, I bundled my family into our rocket ship and flew away.

We stayed briefly at the Planet of Shoes, but found the same problems there. We're happy to have finally found a home where virtue is not founded on attire.

The Joke

Stop me if you've heard this one before. A man goes to the doctor and is diagnosed with cancer; for months afterward he's treated with the very best, cutting-edge medical care, and while his health occasionally dips into a perilous state, and he's often miserable with pain and the stress of his treatments, his cancer eventually goes into remission and his prognosis is good.

What does he say? "Praise Jesus! My prayers were answered!"

Or how about this one: A plane's engine fails on takeoff. The experienced, well-trained pilot coolly keeps his head, follows tested procedures to maintain control of the aircraft, and wrestles it to a rough but safe emergency landing, averting disaster.

What do the passengers say, and what do the media report? "It's a miracle! Thank God for bringing us down safely!"

Here's another. Wildfires sweep through an area, houses burn, people die. Firefighters struggle day after day, set firebreaks, hose down homes, and fly in helicopters with massive buckets of water scooped from nearby lakes. They finally get the fire under control, just shy of destroying some lucky fellow's house.

What does he say? "The Lord has spared me! He must have some special purpose in mind for me!"

These are the kinds of stories you find in the newspapers every day. There's just something wondrously appealing about being saved from a terrible fate that makes people feel good, which is understandable. But then there's an incomprehensible tendency to assign credit for the rescue to some invisible man in the sky who wasn't even there—at the expense of praising the hardworking, skilled human beings who actually were present and were entirely responsible for the happy outcome. It's also sad to imply that the people who die of cancer or in plane crashes or fires must not have been very special or didn't deserve God's helpful attention.

No one seems to blame the omnipotent sky deity for causing the cancer in the first place, or damaging the engine, or setting the fire; that just happens. Or if God did it, it was to teach us a lesson, and he then provides the means of salvation. God gets a lot of good press, and since God never bothers to actually show up and earn the accolades, religion steals the good reputation of the people doing the good work.

More than once I've tried to explain this dishonest theft of human beings' virtue to believers, and there's a common response. So common, in fact, that I can't bear to listen to it anymore and usually cut them off and tell them I've heard it before, I'm unimpressed, and they should find a different excuse. What they do is tell me a story, and it's always the same damned tiresome fable:

> Once upon a time, there was a pious man who was trapped in his house by a flood. As the waters rose higher and higher, he had to climb up on the roof, where he sat down to pray for God to rescue him.
>
> A little later, his neighbors paddled by in a canoe and offered to take him to high ground. "No, thank you," he said. "I have faith that God will save me." And he prayed some more.
>
> The water rose higher, and a rescue boat motored up to his roof and offered to take him to a warm, dry shelter. "No, thank you," he said. "God will not let me down." And he prayed some more.
>
> His roof was almost submerged and he was perched on the chimney when a Coast Guard helicopter swooped down and dangled a ladder in front of him. "No, thank you!" he shouted over the storm. "God will send a miracle!"
>
> And then he drowned. His spirit appeared before God, and he complained, "I was devout and truly believed in You, and prayed and prayed. Why didn't you save me?"
>
> And God said, "I sent you a canoe, a boat, and a helicopter! What more could you ask for?"

Then the storyteller usually sits back, smug and confident. Obviously, I don't know how God works, or else I'd understand that all those natural causes that I say are more important than some fairy tale are actually evidence for God. God works in mysterious ways, and those ways include science, medicine, and engineering. They appropriate the hard work, the years of training, and the entirely secular reasoning behind the practitioners of those disciplines and hand them over to a lazy invisible man in the sky.

It's exasperating. There's usually no point in arguing further, because you're dealing with the sort of mind that thinks Jesus is a subatomic particle that holds the universe together with the power of his love, and that gravity works by means of little angels keeping things down. So I usually just give them a short story of my own.

> Once upon a time there was an atheist who was trapped in her house by a flood. As the waters rose higher and higher, she had to climb up on the roof, where she hoped for rescue.
>
> A little later, her neighbors paddled by in a canoe and offered to take her to high ground.
>
> She got in the canoe.

See? That's the difference between a godly person and an atheist. Our stories are shorter and don't assume the protagonist is an idiot.

The Great Desecration

In the summer of 2008 a pair of students in Florida attended a Catholic mass held in their student union building. During communion one of them, instead of swallowing the wafer, which would have been the holy thing to do, kept it and walked back to his seat holding it in his hand. It was a simple, harmless act, but others in the congregation were outraged and tried to physically wrest it from him. He fled the scene—and then people called for his immediate punishment and expulsion from the university for his gross act of desecration. It wasn't just a local contretemps, either; Bill Donohue of the Catholic League made it a national matter, using his mighty fax machine and his pulpit, while the pundits of Fox News bayed for the destruction of this young man. Over a mere cracker.

The student had no intention of doing any harm and even returned the undamaged scrap of bad bread to the priest a week or so later—after a week of nonstop harassment and accusations and threats. Even then the threats didn't end but continued under the goad of loudmouthed Catholic zealots.

Personally, I couldn't believe it. No one could be that

fanatically devoted to a small scrap of unleavened bread, could they? I decided to join the fracas, in part to test whether people really were that crazy, and also to draw some of the heat off the two students. I mentioned on my blog that this was a ridiculous situation and suggested that people send me a few communion wafers, so I could demonstrate what intentional desecration looked like.

The response floored me.

At first many people were enthusiastic. I got consecrated communion wafers in the mail, and as I traveled, people would press the little crackers into my hands, fresh from the local mass. I even got one mailed to me from England, with a link to a video on YouTube documenting its acquisition from an official Catholic ceremony. I was urged to do something public with it, specifically to protest Catholic misinformation about contraceptives in Africa. An unexpected amount of resentment toward Catholic policies was being expressed.

Then the backlash began. Bill Donohue issued a series of fatwas against me from the office of the Catholic League, which led in turn to an inundation of e-mail and letters from Catholics all across the country. I tried to keep track but gave up when the volume reached roughly fifteen thousand e-mails by the middle of the frenzy. Let me tell you, gang, that can choke most e-mail software out there, and it tended to drown out a lot of the serious messages that were sent to me.

Then I was in a curious situation, holding a stack of consecrated communion wafers. I was holding Jesus hostage, according to some of my critics; the Catholic League was howling for my head on a stick, or at least that I be summarily fired for my affront to their Lord and Savior. Meanwhile a mob of atheists and disgruntled ex-Catholics were demanding that I do something horrible to the donated crackers. I was stumped. What does one do to desecrate a silly piece of bread, a bit of organic matter that even in the most reverent of situations is fated to be slathered with spit, chewed up, swallowed, and eventually excreted?

To find out what the traditional mode of desecration was, I followed the standard academic routine. I looked it up, wondering what the big deal was about these wafers.

I wonder how many of my Catholic friends have heard of the Fourth Lateran Council of 1215. This was the event at which many of their important dogmas were codified, including the ideas of *Extra Ecclesiam nulla salus:* that the Eucharist was the sacrament that only properly ordained priests of the Catholic church could give, and that the Jews were a pariah people, who could hold no public office, had to pay a special Jew tax for their right to exist, and were required to wear special clothing to distinguish them from Christians. The yellow badge marking the *Juden* was not an invention of the Nazis but a decree by faithful Catholics in the Middle Ages. That's an interesting juxtaposition, that a symbol of Christian uniqueness was formalized at the same time that the church formally decreed the Jews to be inferior and a target of hatred.

That combination was useful. Declare that something cheap, disposable, and common is imbued with magic by the words of a priest, and the trivial becomes a powerful token to inflame the mob. Declare a bit of bread to be the most powerful and desirable object in the world, and then you can pretend that the evil other is scheming to deprive the faithful of it. You can invent stories of Jews and witches taking the communion host to torture, to make Jesus suffer even more, and good Catholics will of course rise in horror to defend their salvation. None of the stories were true, of course—Jews and infidels see no power at all in those little crackers, and the idea that they were obsessing over obtaining a nonsacred, powerless, pointless relic is ludicrous. But, heck, it's easy to make accusations illustrated by cheesy woodcuts of hook-nosed Jews hammering nails into communion wafers and lurid tales of blood-spurting crackers and hosts that pulsed like a beating heart, thereby providing a pretext for massacres.

And Christians really did massacre people over the crackers that I had so cavalierly requested. The first recorded accusation that Jews were stealing and desecrating communion wafers was made in 1243, near Berlin; in what was a typically hysterical reaction for the time, the good Germans of the area killed every Jew in the neighborhood by burning them to death. There were more such events throughout history: in Paris in 1290, in Austria in 1294, in Brussels in 1370; more regional pogroms in Germany in 1298, 1299, 1306, 1337; elsewhere in 1338, 1388, 1401, 1420, 1453, 1478, 1492, 1514, 1558—the list goes on and on. Since the thirteenth century, manias for murdering Jews were frequent, and on many of these occasions, their purported theft of communion wafers was the excuse used to justify wholesale bloody murder of entire communities.

That is the true power of the cracker, this silly symbol of superstition. Fortunately, Catholicism has mellowed with age—the last time a Catholic nation rose up to slaughter its non-Christian citizenry was a whole seventy years ago, after all—but the sentiment still lingers. Catholicism has been actively poisoning the minds of its practitioners with the most amazing bullshit for years, and until recently I had no idea that a significant number of people actually believed this nonsense, or that the hatred was still simmering there, waiting for an opportunity to boil over again.

All the regular readers of my blog have seen the thousands of mindless comments by Catholics, demanding that no harm come to a cracker. My e-mail program was melting down with swarms of insults, threats, pleas, and promises of prayers because I threatened to violate one of their holy crackers. In my years of loud and often inflammatory blogging, it was the most impressive demonstration of mass lunacy I have ever seen.

A typical e-mail message would say something like "I was saddened to hear of your plans to harm our Lord Jesus Christ." You would not believe how many people wrote to me insisting that these horrible little wafers (they look like flattened bits of Styrofoam) are literally pieces of their god, and that this omnipotent being who created the universe can actually be seriously harmed by, as they called me, some third-rate liberal intellectual at a third-rate university. (The diminution of my vast powers is also a common theme.)

Other e-mails cranked up the crazy. Some people accused me of being a Freemason, and lots called me a Jew, which is illuminating given the history of this issue. Strange metaphors are common (apparently, I have an uncircumcised heart, which makes me wonder about what kind of bizarre rituals Catholic infants must go through), and of course there were many claims that I was going to burn in hell forever, and that I was under satanic influence.

Another common theme has been the attempt to transform the desecration of a Catholic symbol into the desecration of an Islamic symbol. Obviously, it's not desecration they find disagreeable—it's the idea that someone would offend their weird sectarian sensibilities. I would never dare criticize Muslim rituals, instead assaulting the precious beliefs of kindly, forbearing Christians who would simply turn the other cheek.

Yeah, right. Catholics won't attack me, but Muslims will. Never mind that the Catholic League demands that I be fired, that thousands of Catholics demand that I be kicked out of the university immediately, and that they send me death threats—both the explicit kind and the vaguely menacing kind. Let's not forget the student who started all this by simply walking back to his seat with a cracker only to face censure and possible expulsion from his university. Those Catholics sure are forbearing and tolerant.

A few of these good Catholics learned that I have family—it was no secret—and leaped to use the opportunity. Since I was not demonstrating any fear of the threats directed against me, well hey, let's try a new target! Another writer thought cowardly warnings against my family might be effective and threatened that vengeance "will be carried out on your child," or made up despicable stories about how my deceased father was so devout (he wasn't) and would be ashamed of me now.

Lies and slanders against my innocent and uninvolved family are not going to convince me of anything other than that religion breeds the most disgustingly vile haters in our country, and that Catholicism fits right in with the rest. I will note, however, that Bill Donohue tried to get the Council on American-Islamic Relations to join his crusade, but I have gotten no e-mails from outraged Muslims, over a span of time in which I've received thousands of Catholic hate-mail messages.

If you want to see the deep danger of religion, you have to read this other common sentiment: "I have respected every religion and every idea for years."

Ah, what a beautiful illustration of the completely open mind—utterly undiscriminating, lacking any criteria for acceptance, blissfully and uncritically according every idea its full respect. But, of course, it's a lie: they don't regard every idea as equally deserving. They clearly consider the atheist idea that the sacraments of their faith are empty foolishness to be an outrage. Rather, what they love is the idea that everyone else must respect their beliefs, no matter what they are, and that any disagreement is an insult. This is exactly the kind of uncritical, unskeptical, nonjudgmental idiocy that all religions seek to promulgate, because they all know that if we tore off the blinders of tradition and artificiality and mindless etiquette, we'd see right through their lies. Respect every idea! Especially mine! And if you find stupid the idea that this cracker is a god, why, you must be disrespectful and no gentleman!

My very favorite letter, though, expressed a remarkable lack of perspective. "Your act," the writer said, "is far more deplorable than Hitlers' Holocaust or the terriorists [*sic*] on 9-11."

This is what we've come to: the plight of one small, sad bit of bread is more precious to some people than the murder of six million people. I'm fond of pointing to this letter as an indicator of just how badly religion can mangle normal ethical priorities and generate minds capable of perpetrating atrocities.

For even deeper inanity, let's not forget the Catholic

blogs! We're talking some serious derangement here. Mark Shea regards cracker abuse as profoundly evil:

> I won't mince words. Myers is an evil man. And as evil men, particularly evil intellectuals, tend to be, he is also a mad man as are most of his acolytes and followers.
>
> Myers and Co. are enmeshed in these lies because they have chosen evil. It is evil—archetypally evil—to desecrate the Eucharist. It's the sort of stuff archetypal bad guys in the movies do. It's completely unnecessary gratuitous evil.

To the Mark Sheas of the world, I would say, It's just a cracker.

I think if I were truly evil, I would have to demand that all my acolytes be celibate, although I'd turn a blind eye to any sexual depravities they might commit. If I wanted to be an evil hypocrite, I'd drape myself in expensive jeweled robes and live in an ornate palace while telling all my followers that poverty is a virtue. If I wanted to commit world-class evil, I'd undermine efforts at family planning by the poor, especially if I could simultaneously enable the spread of deadly diseases. And if I wanted to be so evil that I would commit a devastating crime against the whole of the human race, twisting the minds of children into ignorance and hatred, I would promote the indoctrination of religion in children's upbringing and foment hatred against anyone who dared to speak out in defiance.

I'm sorry to say that I aspire to be only a teeny-tiny bit evil, and that my target is a handful of virtually inedible crackers in my possession. It's not much, and all I can say in my defense is: It's a start. A very small start. I'm going to need lots and lots of people to rise up and follow suit, subjecting old, dishonest institutions of hardened dogma to our chief weapon of ridicule and deris . . . our two weapons of ridicule, derision and laughter . . . no, three weapons of ridicule, derision, laughter, and skeptici . . . oh, never mind. You know what I mean. Get to work.

The result of my plotting and my research was unfortunately anticlimactic. Many people proposed intricate plans for how to do horrible things to the crackers, but I repeat, they're just crackers. I wasn't going to make any major investment of time, money, or effort in treating these dabs of unpleasantness as they deserve, because all they deserve is casual disposal. However, inspired by an old woodcut of Jews stabbing the host—it was claimed that they pierced the crackers to make them bleed, then used the potent blood of Jesus Christ in unholy, wicked Jewish rites—I thought of a simple, quick thing to do: I pierced the crackers with a rusty nail. (I hope Jesus's tetanus shots are up to date.) And then I threw them in the trash, followed by the classic decorative items of trash cans everywhere, old coffee grounds and a banana peel. My apologies to those who hoped for more, but the worst I can do is show my unconcerned contempt.

By the way, I didn't want to single out the crackers, so I nailed them to a few ripped-out pages from the Koran and *The God Delusion*. They are just paper. Nothing must be held sacred. Question everything. God is not great, Jesus is not your lord, you are not disciples of any charismatic prophet. You are all human beings who must make your way through your lives by thinking and learning, and you have the job of advancing humanity's knowledge by winnowing out the errors of past generations and finding a deeper understanding of reality. You will not find wisdom in rituals and sacraments and dogma, which build only self-satisfied ignorance. You can find truth by looking at your world with fresh eyes and a questioning mind.

And it was finished.

Get a Job

There is something heretics do that I never anticipated would generate such fury in the godly. This particular act of heresy was so profane, so vile, so revolting that it has triggered legal action all across the country and expressions of outrage everywhere. The words used will sear your eyeballs and invoke the wrath of God, just as opening the Ark of the Covenant did in that Indiana Jones movie.

Apparently, the worst thing a heretic can do is to put up a sign somewhere suggesting that God might be superfluous. Godless folk have been putting up oppressive signs on billboards and buses that carry horrific messages like "You can be good without god," "A million New Yorkers are good without god. Are you?" and "Don't believe in god? You aren't alone." I've had to conclude that mutilating the body of Jesus in the form of a cracker was extravagant overkill. There are much simpler things one can do that Christians will find intolerably offensive.

It's chilling. Atheists are announcing that they aren't going to riot, rob, and rape but are actually going to be cooperative members of society!

These signs incite furious letters to the editor in local

newspapers. Billboard companies refuse to run them. Bus drivers refuse to drive buses bearing them, and sometimes people sneak out and vandalize them. It's rather amazing. How dare atheists suggest that they can act ethically, too! This goes against generations of demonization!

Now, I'm sure True Christians would not resent good behavior on the part of their godless compatriots, so there's got to be some other reason they're upset about these signs. And I think it's because we're in rough economic times, and they sympathize with God.

Face it, he's old. If he's told he's extra baggage, that people are quite capable of being good without him peeping into their bedrooms and sending Pat Robertson whispered messages, then he might be sacked. At his age, that's no light punishment—nobody is going to hire him, and it's a bit late for retraining. Besides, he's way behind the times on all the high-tech stuff, so what's he going to do? Sell used cars on commission? Stand around and be a Wal-Mart greeter?

No, he's going to go on the dole. He'd have to go stand in line at the welfare office, and you know what he'd find there: Zeus and Odin and Ahura Mazda and Marduk and all those other gods he's been lording it over for a thousand years. They'd all just smile smugly as he joined their queue, and I don't think he could bear it. He's used to getting hosannas and fatted calves and lying back in his recliner and munching on foreskins while watching *The 700 Club*. Can you even imagine how expensive a foreskin habit would be if you don't have people offering you mountains of them? They're tiny. It takes tremendous numbers of Christian and Jewish babies to make even a light snack.

Probably the worst thing you can do to a patriarchal

deity is to impugn his career and his manhood, to suggest that he's a useless old fart and that his life's career of whipping fear into people is a failure. Even I feel a little pity for the cranky old geezer.

I'm still handing him his pink slip, though.

It's So Easy to Be Outraged!

Sacrilege can get people killed. It can cause riots and economic mayhem. People die when their sense of propriety is offended. And whose fault is that?

You know what I'm talking about. Almost all religions have odd proscriptions that their followers take with extraordinary seriousness—they are markers for who belongs in the group and who is an outsider. Violating them, especially knowingly, immediately marks you as the Enemy and justifies taking any action against you.

The Muslim outrage about cartoon portrayals of their prophet in 2005 was a recent example. It's utterly ridiculous. Sure, I can see that immigrant populations might be extremely sensitive to further marginalization and might see disrespectful stereotyping as a sign of deeper conflicts. But the response, rioting in the streets, was excessive and irrational and destroyed any possibility that a legitimate grievance might be taken seriously. If you're rightly upset that your ethnicity is being associated with terrorist bombings, it is insane to react by throwing bombs.

But ultimately, suppressing portrayals of the prophet was not just about protecting people from discrimination; it was about enforcing a body of idiosyncratic, peculiar laws as if they were representative of broader human concerns and rights. They aren't. They're just silly.

Denunciation of that sacrilege was also about defending practices sanctified by patriarchal religious beliefs that, to anyone with a civilized, grown-up appreciation of a universal morality, are odious. Morality cannot privilege one ethnic group, one religious belief, one sex—a rational morality has to be constructed around an appreciation of all human beings' rights to justice and equal treatment. Many of the portrayals of the prophet Muhammad are intended to openly disrespect specific practices that are endorsed (or at least uncritically accepted) by members of the Muslim faith but that violate the rights of a much larger subset of the human race.

One of the serious purposes of sacrilege is to tell people that their beliefs are barbaric or wrong. One of the purposes of blasphemy laws is to silence critics of long-hallowed discrimination—like patriarchal domination, where people of one sex are informed over and over again that they are in charge and must be given special favor, and people of the other are told over and over that they are weak and must submit and obey. The Abrahamic religions are rife with this evil myth, and we have to be free to challenge it.

Another purpose of sacrilege is to assert our freedom from superstition. There's a proscription against portraits of Muhammad? How absurd. We defy such arbitrary restrictions on our freedom, be it a demand that we treat crackers with respect or a demand that we do not render images of some guy. We violate these restrictions because we can. Any of us can pick up a Sharpie and scrawl out a picture of Muhammad. Go ahead. Do it yourself.

If you've got some artistic talent, maybe you drew a sketch of a bearded man. If you're like me and aren't particularly skilled, you drew a stick figure or something equally simplistic. Is it actually a picture of Muhammad? How can you tell? Only because you say so. Is it offensive? Only if you choose to regard a crude sketch of a bearded man in a hat as somehow insulting everyone who believes in a particular holy book.

Remember, people will fight and die because they want to stop you from doing something so trivial and harmless as sketching what you think Muhammad looked like. That's stupid and wrong, and it is their problem, not yours. Sacrilege is exercising your freedom, a freedom they don't want you to have.

I Am Not a Spoiled Child Having a Temper Tantrum

Some have accused me of lacking empathy for believers; that's not quite right. When the Florida preacher Terry Jones said he was planning to burn the Koran on September 11, 2010, which led to an Islamic uproar, my response was to say that that's too bad, we have a right to destroy our own property, and that people's reactions to this trivial event were inappropriately hysterical.

Informing me that Muslims are genuinely and sincerely and deeply offended is not informative. Contrary to the suggestion that I must have an empathy deficit, I know and appreciate the fact that their feelings are hurt and that they are angry and outraged. My point is that I don't care, and no one else has to be angry, either. The Abrahamic religions—Christianity, Islam, and Judaism—are all about fostering a feeling of oppression as a tool to reinforce tribalism, and hearing yet another deranged member of the People of the Book whine that we show insufficient respect for their mythology exasperates me the same way my small children wailing about not getting a candy bar in the grocery store does. Fine, you can be mad all you want, but that doesn't obligate me to cater to your whims.

Islam has been particularly petty and overly sensitive to slights, and it seems to be even more narrow and provincial than other religions in its lack of awareness that the majority of the people on the planet do not hold its beliefs in any great reverence and that Muslims don't get to demand that we treat their superstitions as sacred. Their leadership seems to have decided that they do have the right to enforce Muslim piety on non-Muslims, rather than recognizing their own minority status and appreciating—and perhaps, dare I say it, empathizing with—people who hold very different beliefs. As CNS News has reported,

> Following the uproar over the threatened burning of the Quran by a small Florida church, a leading international Islamic body said Thursday that the United Nations should outlaw "all forms of offense against religions." . . . "It is a blot on humanity that such discriminatory attack against Islam and Islamic holy sites is continuing in the absence of deterrent legal measures, local and international."

These intolerant thugs for Allah want to criminalize any criticisms of their beliefs and practices. They are mere goons who want to force everyone to respect their peculiar faith.

I'm not joking when I say they want to criminalize free speech.

In Iran, senior Shi'ite cleric Grand Ayatollah Saafi Gulpaygani said Terry Jones should be arrested immediately and his church permanently shut down.

Imagine that.

I've seen enough videos of Middle Eastern protesters setting American flags on fire that I do have to wonder—how would they feel if Americans asserted that people who disrespect the United States ought to be arrested and their demonstrations shut down by force? The people Jones has offended feel no compunction about offending Americans—and that's okay. It's not a crime to offend others; in fact, it's pretty much a natural consequence of having diverse cultures. We live in a world full of people with different backgrounds, interests, and values, and we must learn to accept that there will be clashes. As long as the differences do no harm, we should back off and accept them.

Grown-ups compensate for differences among people that cause no physical harm. We follow a principle of tolerance, of real tolerance, by permitting others to do as they will, as long as it doesn't infringe on the freedom of others. Being offended or desiring to impose your morality on others is not an excuse to violate this principle. Why is this so hard for some people to understand?

Swaddling somebody else's stuff in "meaning" doesn't justify demanding that they treat it the way you treat yours. I had a couple of Bibles in my house; to me, they're just reference books, and not very good ones—I often prefer to pull up a searchable online version to digging into my shelves. These same books may be sacred objects to someone else; others may invest their precious Bibles with significance and reverence; but that doesn't mean I am obligated to regard mine with the same perspective. If you tell me, "My Bible/Koran/cracker/icon must be treated with respect," I'll agree—it's yours. I'll respect your right to revere it however you want. But nobody gets to dictate to me how I should treat my Bible/Koran/cracker/icon—I get to do as I will with it.

Peter Lipson drew a bizarre analogy that makes my point for me.

> Whether or not you think it appropriate, people imbue objects with meaning. Why else try to save your house from burning down? You have insurance, don't you? But most people don't want to lose a house and the objects it contains because they have meaning. Religious objects are no more or less irrationally revered than family photos. People give them meaning.

Errm, my house is my house—I live in it. Burning it down would cause me harm and discomfort. My family photos are my family photos; they do have meaning to me. But the so-called holy books on my shelves are trivially replaceable, and I'd leave them behind in a fire in a minute. Why should I risk my life to run in and rescue them because some fundy somewhere thinks they're precious? Should I grab the Bible off the shelf instead of the family photo album? What right do others have to rank the importance of my possessions to me?

This concept seems difficult for some people to grasp. No one is saying you can't irrationally revere some religious object—you just can't tell others that they must irrationally revere your religious object, and you especially can't tell others that they must treat their cheap, mass-produced copy of your religious object in some special way.

One final point: the irrational adoration of religious trash does people harm. The marching Muslims howling about some nobody in Florida threatening to burn copies of their book bring to mind Thomas Frank's *What's the Matter with Kansas?*. These are people who have been manipulated by the lies of faith, who have been duped into thinking the most important determinant of their happiness and status in life is their dedication to religious abstractions—and it leads them to work against their own interests.

The Middle East is a region rich in mineral wealth, but that wealth is being used to prop up a medieval hierarchy or a corrupt dictatorship/monarchy or a theocracy. It should be rich and prosperous, with a growing middle class and declining poverty. Without the handicap of its religion and historical traditions, the region should be more like Norway economically. But it's wracked with wars and the relics of colonialism. The West is still barging in militarily and causing devastation. Muslims in those countries should be righteously pissed off, but not about something as trivial as copies of their favorite book being destroyed.

That's empathy, too—the awareness that Muslims are human beings who deserve better, and that getting distracted by such pointless noise is doing them harm. (I should also add that the Muslim world also protests rightfully and vociferously against the military atrocities committed against them. I think much of the news about the Koran-burning protests overhypes the reaction, in part to make the Islamic world look ridiculous. It's a distraction on both sides.)

Religion infantilizes people. It makes them humorless and blind to others' ideas. We're doing no favor to religious people by indulging their unrealistic and impossible dreams of controlling everyone else's life.

The Proper Fate for a Holy Book

Lately, considerable angst and fury have erupted over a bad book, the Koran. Terry Jones, a fundamentalist lackwit, announces that he's going to set his copy on fire, and he calls on his fellow travelers in extreme Christianity to join him so they can have a regular bonfire. Let's burn lots of books! This has been a common practice to inflame anger: have a book burning, or a flag burning, or a rock-and-roll conflagration.

I protest. I detest the Koran as well, and also the Bible and every other holy book out there, but burning the target of your hatred is a very bad idea. Don't they know about CO_2-driven climate change? Books are made of cellulose, which contains lots of interlocked long-chain carbon molecules, and burning them simply releases all that carbon into the air, where it contributes in a small way to anthropogenic global warming. Stop it! Instead, try carbon sequestration: bury the books.

I know, I know, it doesn't quite have the panache or generate the spectacle of flames licking upward into the sky, and it's going to leave the pyromaniacs in your movement frustrated, but think about the PR advantages. You dig a deep hole, you throw what you hate into it—be it the Koran or the Bible or your enemy's flag or those Elvis and Beatles records—you cover it up with topsoil, and then you plant flowers on top of it. That's right, turn anathema into flowers. Everyone loves flowers. And the plants take more carbon out of the air and release more oxygen! It's win-win-win-win, a triumph every way you look at it.

At my house, I have a surplus of Bibles and Korans. People keep sending them to me, as if it will help, and I can't give them to the local library because they really only need one copy of each (and I doubt that they're very popular for checkout), so they accumulate. I don't mind having the books around, although it's actually superfluous since the copies on my computer are much handier to access and search, but my bookshelves are groaning.

So I did the sensible thing and buried some in my backyard, and I planted pansies on top of them.

It's not only practical; it's deeply and personally satisfying. Burying may not be as fast and flashy as setting a fire, but if you're aware of what will happen over time, it's an even ruder expression of contempt for the foolish ideas that those bad books contain.

Imagine how the pages will swell with moisture, the fibers will separate, and the chapters will turn into pulpy masses. Bacteria will bloom and their colonies expand; fungi will flourish and their hyphae will infiltrate and convert cellulose into spores. The ink will run, as nematodes writhe over the surfaces, etching the words with slime and replacing the follies of dead men with the wisdom of worms. The roots of flowers and grasses will fumble downward to embrace the decaying leaves, and the roots of trees will impale the volumes laterally. Given only a little time, the madness will be reduced to compost.

At every instant in this gradual process of degradation, the books are being improved and given greater value. And my decision to discard the poisonous symbols of past ignorance makes me a little more free. It feels so good.

Ask but Don't Tell

I've been following the news lately and have at last unearthed the most horrible, awful, evil thing you can do to a religion, one simple thing that will get the faithful to melt down.

Tattling.

Thou shalt not tell on the church! It ought to be the first commandment. Church leaders can engage in the most ghastly, demeaning, terrible crimes, like raping children, but the believers' concern isn't for the young people who've been hurt. Instead they worry that the revelation of human imperfection among priests might diminish people's dedication to the faith, so it must be covered up. The guilty can't be punished, because that would be an admission of failure by the hierarchy—so they get shuttled off to some new parish, their crimes concealed.

Most civilized people would agree that child molestation is one of the worst crimes—it's a betrayal of our relationship with children. It's damaging psychologically and physically.

The Catholic Church seems to think otherwise. Raping

children is something to be tut-tutted over, the miscreant quietly shuffled off to some safe haven. Scorn is reserved for those who tattle on them.

Bill Donohue, the ever-reliable paladin of the Catholic Church, responded to accusations against a priest who had abused more than two hundred children in Wisconsin by finding fault with the real problem: the families! "I do blame the families," he declared. They'd taken decades to overcome their fear and respect for Catholic dominion to finally levy charges, and to the devout, this was the true crime. A monster cloaked in godly authority takes advantage of an entire community, and Donohue's words of condemnation are entirely for the families who were afraid to speak up for so many years.

Well, they're speaking up now, Bill. They're defying heavenly authority, no doubt torn because they've been told all their lives that their immortal souls are dependent on the intervention and aid of the man who has been sticking his penis in their sons and daughters, and you're just confirming what they already knew: that testifying would only get them blame and condemnation from the ardent moralists of the church. *Thou shalt not tattle, but if you do tattle, we'll piss on you for not tattling soon enough.*

That's a mighty fine catch, that catch-22.

Another excuse they use is "Everyone does it." You'll find no more repulsive example than a rationalization by Catholic Archbishop Timothy M. Dolan of New York—his excuse is to list examples of doctors, judges, and teachers who abuse children. See? Catholics probably aren't any worse than other professionals! Except, of course, that medicine, law, and education aren't closed institutions that go out of their way to protect molesters from exposure. And most of all, they aren't institutions that insist, over and over, that they are repositories of moral wisdom.

That particularly galls us atheists, who are so tired of being told that we can't be moral without an objective external source of goodness. Yet here we have a group of people who most loudly claim as their professional calling a direct insight into the mind and will of the cosmic lawgiver—and what do we find? Violations of basic human decency at a most profound level.

And the archbishop's excuse reveals another fundamental moral failing. What kind of child thinks that is a legitimate excuse? Matt Taibbi puts it well:

> But even worse—what does Dolan's whiny deflecting and excuse-making say about the church as an arbiter of ethical values? These pompous assholes run around in their poofy robes and dresses shaking smoke-filled decanters with important expressions on their faces and pretending to great insight about grace and humility, but here we have the head of the largest diocese in America teaching his entire congregation that when caught committing a terrible sin, the appropriate response is to blame the media and pull the "All the other kids were doing it, too!" stunt!

Here's another excuse. We atheists have been blamed for creationism—literally told that our outspoken godlessness drives people of faith to become creationists—by the likes of Ken Miller, so you just knew that someone with an even more exalted status in the church would find us a convenient scapegoat for even more serious problems. Pope Benedict had an explanation for the atrocities in Ireland: "new and serious challenges to the faith arising from the rapid transformation and secularization of Irish society."

A priest buggered a little boy? Look—an atheist! It's his fault! How dare he stir up disturbing changes in society? Father Murphy was so distressed, he had to stick his penis in Timmy's butt! Aww, poor Father Murphy . . .

Some people are waking up. Even in Pope Benedict's hometown of Traunstein, people are disillusioned.

> "We were proud to have a German Pope," one woman tells me, "but that pride has gone."
>
> "The abuse scandal comes as a shock," says another. "Now people are leaving the Church."

I have no confidence that the effect will be lasting—people seem to be very good at putting aside their real-world concerns when someone with pretensions to religious authority tells them God wants them to. But most of all—and here's the difference between Christian and atheist morality—I would not want to see even the complete dissolution of their odious faith if it could be achieved with the pain of a single child. We do not build our world on scapegoats and blood sacrifice and a belief that torture and death are redeeming. But the religious do. They can justify turning a blind eye to tormented children because it is a small sacrifice to make to an omnipotent god; they have saints who glorified poverty and misery; they'll tell Africans to expose themselves to deadly disease because prophylactics are a sin.

There is a wonderful story by Ursula K. Le Guin, "The Ones Who Walk Away from Omelas," which is a fantasy tale about a beautiful city called Omelas, with wonderful people and a thriving, joyous culture, but one catch: the city is built on one necessary case of deep suffering. There is one small child locked away in a filthy basement, deprived of human comfort, left in "abominable misery," and the prosperity of the new city is magically coupled to the pain of an innocent child. Most people accept this price and close their minds to it, dealing with it by ignoring it.

In the story, a few people reject the terms of the sacrifice, and this is where I have mixed feelings about it: they quietly leave the city, alone, to go somewhere else, somewhere they can't even imagine. They are "the ones who walk away from Omelas," after all. They are the ones who will not accept bliss if it's founded on another's pain.

But I don't know; it doesn't seem enough. The story seems to accept that there is an act that cannot be committed: that protest and tattling are not an option, that no one ever sees that there is justice in bringing the oppressed to light and doing something. Or even questioning whether a child's sacrifice is at all causal in bringing about their happiness. It's a thoughtful story, but it needs a sequel, "The Ones Who Storm the Gates of Omelas."

Dirty Words

Words are the great juju—some apparently believe we have the power to call up Satan and summon the lightning with our use of language. One of the common quirks of many Christian and Jewish writings is the insistence on writing G_D, as if including an *o* would turn the word into a Rune of Power, an expression of disrespect, or instill some strange fear in the writer. It's God as Voldemort, and all I can say is F_CK THAT.

> There ought to be a room in every house to swear in. It's dangerous to have to repress an emotion like that.
>
> MARK TWAIN

God damn it.

I was brought up to think that that phrase was a great sin—gentlemen simply didn't say that. And according to an especially patronizing attitude, one particularly didn't say anything like that in front of women. Apparently strong words were too much for the weaker sex.

And that's what it was all about: strong words. Words that confront and challenge, words that are expressive and exuberant, words that defy gods and authority, lusty words that stir up primal urges, violent words that hit hard, even when no physical assault is likely or even possible. Fightin' words. Profane words. Words that you can't say in a church but can shout in a saloon.

Language is power, and some people want to shackle it and tame it and tell you what you are allowed to say—and they especially want to limit words that might be an affront to people who hold an unquestioned faith or to people in power. Swearing is low-class and vulgar. The religious want to take the power of language and turn it into an object of shame. Language is free, after all, and the lower classes and women can use that power easily—and we can't have that. Chain them down with shame, and weaken one weapon that anyone can use.

> My swearing doesn't mean any more to me than your sermons do to you.
>
> MARK TWAIN

But ultimately, words have no magic. There are no gods who can damn anything for us, and merely cussin' someone out does them no physical harm. Profanity is expression; it's a flavor of words, like poetry or song. They all have impact in their own way, but we should no more condemn an obscenity shouted in frustration than we would poetry whispered in love.

> Under certain circumstances, urgent circumstances, desperate circumstances, profanity provides a relief denied even to prayer.
>
> MARK TWAIN

What matters isn't the words, it's the meaning. The words are there to convey significance, and it's dishonest to ourselves to pretend, to mask our meaning by declaring the relevant portion of our vocabulary off-limits. Sometimes a heartfelt *fuck you* is exactly the sentiment we want to express; why should we stitch our lips shut and hide our feelings? Because someone else is incapable of coping with honest emotion?

> The spirit of wrath and not the words—is the sin; and the spirit of wrath is cursing. We begin to swear before we can talk.
>
> MARK TWAIN

Words do have power and can hit us hard, but it is ridiculous to pretend that it is the words alone that are hurtful, that somehow by screening out a few strong words we can shelter tender ears and delicate minds. The most hurtful words I can think of, the phrase that can cause a great amount of actual pain, are *I don't love you anymore.* There's not one obscenity in there. Shall we censor talk of love because it might make someone weep?

> The idea that no gentleman ever swears is all wrong. He can swear and still be a gentleman if he does it in a nice and benevolent and affectionate way.
>
> MARK TWAIN

It's all about how the words are used, not the words themselves. People who are hung up on tiny fragments of language, on single words and phrases, and who are intimidated or angered by them while ignoring the whole of the conversation, are shallow thinkers, superficial censors with no interest in the ideas, only in channeling a conversation along narrow avenues that will not burst out beyond the circumscribed boundaries of their sense of propriety.

The only sin in profanity, to my ears, is when it reflects a lack of imagination. If you're saying *fuck* or *damn* as every other word, you're doing it wrong. But you aren't making Jesus angry, you're just making me bored.

The Top Ten Reasons Religion Is Like Pornography

1. It has been practiced for all of human history, in all cultures.
2. It exploits perfectly natural, even commendable, impulses.
3. Its virtues are debatable, its proponents fanatical.
4. People love it but can't give a rational reason for it.
5. It objectifies and degrades women while it worships them.
6. You want to wash up after shaking hands with any of its producers.
7. The costumes are outrageous, the performances silly, the plots unbelievable.
8. There's nothing wrong with enjoying it, but it's nothing to be proud of, either.
9. It is not a sound basis for public policy, government, or international relations—but it's much more interesting.
10. Its stars are totally fake.

The Purpose-Free Life

I don't want to live a purpose-driven life. You'd have to be insane to aspire to a life defined by someone else—especially if that someone else was a megalomaniacal alien.

Rick Warren's book *The Purpose-Driven Life* is a publishing phenomenon, having sold over 30 million copies, earning its author tens of millions of dollars, and driving Warren straight into the media and public eye, where he's landed influential gigs grilling presidential candidates and advising politicians. The book has to be read, if only to figure out what all the noise is about. So I read it, and I was as appalled as I thought I would be. Do people actually find inspiration in the story Warren tells?

Obviously, the big question the book has to answer lies in the title. The short version of the slender book is this: We exist to serve an immensely powerful extraterrestrial intelligence that demands our unquestioning obedience and our lifelong dedication to worshipping it, and we must also propagate this message of servitude by enlisting more of our fellow earthlings in the cause. Warren says this demanding being's name is God, or sometimes Jesus, but it might just as well be Emperor Ming the Merciless, since he's an uncompromising cosmic tyrant with the most elaborate torture chamber in the universe. Warren should have given the book a more explicit and honest title, like *You Are But a Slave, Worm!*. Oh, but perhaps the situation isn't quite so awful—this tyrant does at least say he loves you, as long as you grovel and bring in your quota of converts.

The first section of the book is titled "You Were Planned for God's Pleasure" and explains that the reason we were all created was to worship this peculiarly vain being; the rest discusses how to build cells of like-minded slaves who will work together on missions to convert other people, and insists that everyone will find fulfillment in such servile goals. It couldn't be more chilling if it were an instruction manual for spreading an infection of mind-control slugs from Ceti Alpha V.

The book's faulty premise is laid out in the first few pages.

> The search for the purpose of life has puzzled people for thousands of years. That's because we typically begin at the wrong starting point—ourselves. . . . You were made for God, not vice versa, and life is about letting God use you for his purposes, not your using him for your own purposes.

To quote Wolfgang Pauli, "That's not right. That's not even wrong." I agree that the search for life's purpose puzzles many people, including many very intelligent people, but then the search for the Philosopher's Stone puzzled many people, as did the search for El Dorado, Shangri-La,

Atlantis, and Santa Claus's Workshop. (The latter, admittedly, baffles only brilliant people under the age of five.) There's a simple explanation for why these mysteries eluded them: they don't exist.

Apply Occam's razor to this situation. Millions or billions of people have been searching for the meaning of life over the course of millennia, including some of the cleverest philosophers who have ever lived. They've found no answer and continue to be puzzled; or they've found a wide range of different answers with no consensus, which reflects the most likely truth, that there is no simple purpose to find. Then suppose an unexceptional Baptist preacher in California with a divinity degree happens to discover that there is one grand purpose, and that it fortuitously coincides with the doctrines of his religious upbringing. Why should we expect that the failed methodology of generations of priests will suddenly bear fruit now, in this one specific sect, in this one specific person? After all, I'm sure Aztec priests five hundred years ago were unassailably confident that the purpose of life was to rip the beating hearts out of prisoners as an offering to keep the sun rising every day. While Rick Warren is significantly less violent, his purposes have just about as much justification.

This is another notable fact of this weirdly popular book: Warren never explains how he knows what he claims to know. If a used-car salesman told you that you must buy the 1991 Chevy sedan on his lot, you'd at least expect him to explain what was so great about that particular car and why it is suitable for you. But Warren simply launches into his flat assertion that you *will* obey Ming—I mean, Jesus. Any reasonably skeptical thinker ought to at least request some reason to obey the invisible man of the Christian religion before groveling, I should hope, and any reasonable proponent of a specific faith ought to be able to provide some verifiable reason to grovel. Warren just assumes.

But as an atheist, I'm here to give you a message of liberation. There is no pan-galactic overlord reading your mind and telling you what to do. Or, at least, there is no evidence that such a being exists, has ever existed, or has any effect on the world. Rejoice!

Furthermore, there is good reason to believe positively in an absence of larger purpose. The history of life on earth is one of fits and starts, expansion and retraction, odd detours, and unpredictable tangents. No species is or ever has been privileged by a deity; we're all of us, from ants to anteaters, subject to disease, predation, death, and extinction. If a god's deepest desire was to create a race of worshipful servants, he took a ridiculously circuitous route to building one, dithering about for 9 billion years before creating the earth, and 4.5 billion years after that before creating human beings, and of that one species, waiting until 3,000 or 4,000 years ago to inform a tiny fraction of the population that they ought to snip off the ends of their penises to properly serve the Lord. Then he dawdled another 1,000 or so years before informing another subset that they were supposed to worship his son, Jesus, and those believers wallowed in ignorance for another 1,500 years before figuring out that Protestantism was the proper path to follow, and another 350 years before deciding that the Southern Baptist sect was the one true religion, and only in this one brief lifetime do we get the final word on our purpose from Rick Warren.

God seems to have some serious communication problems.

More likely, nothing is issuing any orders. Life certainly behaves without apparent commands from above; you are free to worship Vishnu if you like, or no one at all, and you'll receive no angry memos from heaven, bring on yourself no thunderbolts of retribution, and no dock in salary from on high.

There are no rewards, no afterlife. The universe doesn't care what you do. The only consequences of your actions are whether your genes and your ideas are viable and propagate into later generations, and even at that, there is no supervising being who cares whether you propagate or not; the next generation will make do with whatever individuals contribute to it, and you won't be noticed if you don't.

Perhaps that sounds bleak to you, but it shouldn't: it's a message of complete freedom. You are not a slave to any superbeing, and you have all the responsibilities and privileges of autonomy. However, you do have some predispositions. You are the product of a long line of organisms who have all left positive contributions to their succeeding generations, and who have shaped their environment to make it more congenial to their existence with such inventions as culture and technology, and naturally enough, such institutions are also congenial to your existence. You can choose to thrive within them and add to them, or you can ignore your opportunities and responsibilities and die.

It's really that simple. No one is above you, telling you what to do. All of the "purpose" is within you and nowhere else. You've been born into a world where previous generations have left you a toolbox that they used to build their lives, and the next generation will simply use whatever you leave them, if anything, and they will find their own purposes, with no sense of obligation to yours. It's beautiful. It's open.

No purpose is the best kind of purpose there is.

Happy Easter!

On Easter, the masses gawp in awe at a bizarre and unbelievable story—because it is such a good example of how religion piggybacks on our cognitive biases.

You all know the Easter story: a god turns into a man, gets tortured and killed, rises from the dead, and, somehow, this act makes us all better. It's a tale best left unexamined, because it makes no sense. We are supposed to wallow in an emotional thrill that taps deep into our social consciousness, not think about what the story actually says.

The part of the story that works for us is the idea of self-sacrifice. That's potent; we're social animals, and an individual sacrificing him- or herself for the greater good has a lot of impact, materially and symbolically, and stirs up powerful and conflicting emotions. Think about a real example—a soldier throwing himself on a hand grenade to protect his compatriots. It's a noble sacrifice, meaning that one dies so that others may live. It makes us wonder whether we would be brave enough (or crazy enough, or despairing enough) to do the same. We regard someone who does that as a genuine hero, someone who cared so much for his fellow human beings that he would make the supreme sacrifice to spare them.

So that's the aspect of the Easter story that the Christian faith milks for everything it's worth. It amplifies the suffering of Jesus: look at Catholics' weird obsession with graphic portrayals of the bloody, twisted, tormented Christ on a stick; look at Mel Gibson's horrible 2004 torture-porn movie, *The Passion of the Christ,* that lingers sickly over every lash of the whip, every beating, the long slow bleeding death. This isn't just a quick self-sacrifice—Jesus suffered a long lingering death, just for you. He must have cared about you so much!

Uh, except for one thing. Where's the grenade? What is he saving us from?

This is where the myth falls apart. Don't look! Be distracted by the crown of thorns and the spear and the nails, and by the magic trick on the third day! Whatever you do, don't question the sacrifice!

Because, unfortunately, Jesus didn't save us from anything real, and his death made no change in the world. Ask a Christian, and they'll tell you he's saving us from original sin, our flawed, weak, inherently wicked natures. But that sin was an act committed by a pair of mythological ancestors (they didn't even actually exist), and the sin was to be willful, curious, and disobedient to an imaginary man in the sky—it was a nonexistent crime. I don't believe in being held accountable for my ancestors' weaknesses (as Patti Smith sang, "Jesus died for somebody's sins, but not mine"), and in this case I don't even consider what they did to be wrong. So Jesus suffered to atone for an act that I would consider a virtue, committed by myths against a myth? That's no hand grenade, that's a fairy tale. Nobody needs to die to protect me from a fairy tale.

Next problem: What Jesus did doesn't even protect me from that fairy tale! Imagine that in some metaphorical sense, some heritable taint truly did infect the entire human race, passed from generation to generation, making us more prone to do wickedness.

How does it make us better and healthier for the sick, for us fallen people infected with sin, to butcher Doctor Jesus?

I can guess how Christians will rationalize it: that having the doctor make such a sacrifice will make us believe more sincerely in his prescription. But again, that's religion leeching off a cognitive shortcut that our brains take: we'll assume it must have been a very, very important message if the messenger was willing to die for it. That's invalid—people die stupidly for bad reasons all the time. The only test that matters is whether the doctor actually helps people with his actions.

Another problem: Jesus cheats. We're supposed to believe that he's saving us from an imaginary ancestral sin, and that he's doing so by dying—but he doesn't! He comes back three days (okay, actually a day and a half) later, perfectly healthy except for a few holes that don't seem to perturb him much, and he gets to magically zoom up into the sky and live forever in his dad's palace. This is no sacrifice at all.

Now, suppose our hypothetical soldier who threw himself on a grenade turned out to survive the experience hale and healthy because the bomb was a dud. He'd still be a hero—he didn't know it would fizzle, and the intent was there. This doesn't help Jesus, though. He's omnipotent and omniscient and knew his own nature, and he knew that you don't kill a god by hanging him from a tree and poking him with sticks. Jesus faked his heroism. He's no hero at all.

Finally, there's a layer of the story that doesn't resonate with modern audiences much at all: the idea of the propitiatory sacrifice. Got a problem that you think is caused by a god? Take your strong king or your beautiful virgin and kill them, give them to the god, so he'll reward you with a good harvest or good fortune in war or the return of tasty game animals. Most of us know this won't work. Even Christianity tends to steer clear of this claim nowadays, although it is sprinkled throughout the Old Testament. But on another level, the Easter story is the tale of God giving his only-begotten son in a blood sacrifice to propitiate himself and grant us forgiveness for having crossed him once in 4004 BC.

It doesn't work logically or emotionally. It's the action of a psychopath with a grudge over a petty slight; it's what a demented monster would do. We don't regard as heroic the soldier who throws the fellow next to him on top of the grenade, and we especially condemn the soldier who pulls the pin on the grenade, then smothers the explosion with his bunkmate's body.

This is how you should think about the Easter holiday today. Selfless sacrifice for the greater good is a social virtue, and it's nice that Christianity has embedded the idea in its heart. But what's revolting is that Christians take this simple idea and spindle it up into a sick, twisted, confusing botch of a story that guts it of its power if you use your brain and think about it. Easter is a holiday for the mindless, with a grim horror at its center. Even the bunnies know better.

Afterlife? What Afterlife?

Religion flourishes in ignorance. It's a collection of stories and claims that purport to explain the mysteries of life. Whenever people don't understand something or lack real knowledge about it, there is a priest ready to rush in and fill the gap with a story. And it's always a story that provides the answer people want to hear. It's all about retribution for the wicked and rewards for the godly, and everything has a purpose, even the most arbitrary phenomena, because people love to believe in a guiding hand of a god who, if they properly satisfy him, will give them an extra scrap of protection in a dangerous world.

Look at the stories religious people make up. They know nothing of the deep history of the world, so they create a fable about a humanlike deity as one would construct a house. They know nothing of disease, so they make up imaginary demons and spirits that torment us. They don't understand geology or the weather, so every natural catastrophe becomes a warning shot from an angry god. They want power, so they pretend that their incantations and rituals will get them the blessings of their god. Most pathetic of all, they fear death, so they've invented fabulous heavens to tempt and hells to terrify.

They're all lies. Religious people don't know what happens after we die—they can only pretend to know. No one has returned from an unambiguous death to tell us what goes on afterward, and the people who claim to have had visions of an afterlife or ghosts or souls are not consistent with one another. The only reason to invent a story that we have a "spirit" that will "live on" after death is that it's what people want to hear. Death is frightening, so it's easy for people to believe in an afterlife.

But there is no evidence at all for an afterlife. We have no logical reason to believe in it; immortality doesn't even make sense, since a life without growth and change would be no life at all, and an eternity of continuing change could only render who we are now rather irrelevant. The available evidence suggests rather strongly that our minds are dynamic processes that play out on the substrate of our brains (and a theologian wishing otherwise is no rebuttal), and when the activity stops, we cease to exist. I am a unique array of synapses that have been tuned by my personal experience and neural connections laid down under the dictates of genes and development, and when my brain stops and rots, all those memories, every detail of my personality, everything about my mind, will be gone forever.

To argue otherwise is pure fantasy. (At least that's what logic dictates.)

It's a hard sell for atheists, isn't it? While our opponents promise paradise, atheists offer nothing but the prospect of personal oblivion. If all people had to go on was belief, they'd have to be crazy to go with the atheists. But we do have something more than belief: we have reason and evidence and, most important of all, an overriding interest in knowing the truth. Why, we'll accept the most horrible, terrifying ideas if they are true: that we'll fall if we jump off a ten-story building, that we can electrocute ourselves if we stick a piece of silverware into an electric socket, and that someday we will inevitably stop and no longer exist.

Reality matters. The only way to argue for an afterlife is to believe otherwise, to believe that what exists is unimportant compared to what you wish existed. I can't do that. I can't offer anyone soothing words and the promise of consolation, because I have none. We stand naked before the universe, a product of its rules, and one of the facts of our existence is our inevitable obliteration. Running away from that fact won't help. Believing in a magical savior won't help. Hiding in fear won't help.

The essential principle, though, the one that the religious cannot abide, is that you can face it honestly.

And there's at least a little dignity in that.

Soulless!

Atheists don't believe in God.

Atheists deny the Holy Spirit.

Jesus was just a man, at best, as were Buddha, Muhammad, and every other prophet and religious figure in history.

That much everyone seems to be able to pick up on. But I think there's something even more important that atheists reject.

Atheists don't believe in souls!

Now that's a heresy, and it should be even more distressing to people than our denial of gods. As humans we have no immortal, constant part that will survive after death—our minds are the product of a material brain. Atheists understand that they are soulless machines made of meat, honed by millions of years of ruthless, pitiless evolution. And that everyone else is, too.

When we die, we will enter no paradise, no hell, not even a grim gray afterlife of darkness and regret—we will just be gone. Everyone who has ever lived has simply ended and become nonexistent.

That should be the scariest, most depressing part about being an atheist. We have no future? How awful.

But it isn't.

I have several replies to believers who think we should be miserable because we don't have an afterlife to look forward to.

One answer is that lies are not reassuring at all. Telling me that I'll go to heaven when I die is about as believable as telling me I'll be rewarded with beer volcanoes and strippers for my irreverence. I'd rather be honest and aware than deluded and dumb.

Another answer is that we are alive right now—we simply do not have to worry about what will happen after we're dead. Life is for living, and it's wasteful to spend it fretting over what you'll do when it's gone. One reasonable response to mortality is to enjoy life now.

Still another answer is that we do have hope for the future. Think for a moment about your community a century from now. Does it make you feel good to think that there will still be people living there then? That they'll be talking about things that you find interesting and doing activities you also enjoy? Do you hope that life will be better for them? Even though we will be gone, we can still aspire to perpetuate our culture and find satisfaction that we are advancing that cause while we are alive.

The hardest explanation for believers to grasp, though, is the fact that none of us has ever had this unlikely clot of vapor called a soul. If the soul is an imaginary fantasy, then Mozart's music, Michelangelo's sculptures, Picasso's paintings, the Wright brothers' plane, every work of art and technology ever produced by people whose names have been lost to us, every child, every dream, has been created by us, mere mortal flesh unled by a magic puppeteer in the sky, unaided by angels or spirits. I find that wonderful.

We atheists also understand that no purpose has been bestowed upon us by an imaginary creator, and no task is to be executed by our busy purposeful souls to win us a spot in an imaginary heaven. We don't believe the universe has been granted a grand purpose by some kind of deity—something central to the Abrahamic religions, at least—and we do offend Christians by denying a key tenet of their faith.

And that's just fine. We not only deny it, we're proud of our understanding of reality. It's not just atheists who reject the notion of a god-granted purpose—the nature of the universe repudiates it. So let's be clear: here is an atheist's understanding of our place in the cosmos.

Each of us is like a wave in the ocean, a gust of wind, a lightning bolt—unique, yet at the same time part of a pattern of forces. None of us has any special privilege in the universe.

- Our minds are the product of intricate, interlinked chemical processes. The mind is complex, but it is also fragile, requiring a delicate balance of minute quantities of ions and molecules to function properly.
- We as humans are impermanent. The wave will crash on the shore, the gust of wind will dissipate, the lightning bolt will flash and fade—and they will never return. One day we all will die. We will simply cease to exist.
- We atheists are not afraid. It takes courage to confront and accept reality. The person who has the perspective to appreciate his or her true place in the world, who can reconcile him- or herself to mortality and work forthrightly for the truth, is not the frightened one.

Courage is about recognizing reality and going on anyway.

Atheists are not the ones trying to paper over the abyss with wishful thinking about magical father figures who are somehow invisible and intangible and never make a scrap of material difference in the world. We aren't the ones deluding ourselves with fantasies about achieving immortality as ghosts or getting our wishes granted by angels.

Atheists do not accept the idea of immortality—which means we can also throw away the irrelevant crutches of god-belief. We've walked successfully without them; we've cast them away, stand as proud human beings, and meet fate without the wishful delusions of faith. That's why Christians consider the idea of soullessness a sacrilege—it says that we have no need of priests or gods or persistent ghosthood, ideas that have only hobbled us.

What Dreadful Price Must We Pay to Be Atheists?

It seems that in 2010 Dr. R. Albert Mohler, Jr., the president of the Southern Baptist Theological Seminary, was feuding with Michael Dowd, the author of *Thank God for Evolution,* who endorses a kind of fuzzy spirituality that is mostly pro-science. I can't honestly say I'm a fan of Dowd's approach. (Bite the bullet already, man—there's no need for even the concept of spirituality.) But at least a fuzzy faith is preferable to the cast-iron dogma of an old-school Baptist.

Here's the thing: without even trying, Al Mohler is hilarious. Also, a little scary. He's wonderfully oblivious to what he writes, so his screed against Dowd, damning him for rejecting fundamentalist doctrine to embrace science, gives us this jewel.

> I regret to learn of Michael Dowd's cancer, but my concern for him is far more urgently focused on his malignant beliefs. In his own very effective way, Dowd clarifies the theological and biblical costs of embracing the evolutionary worldview. In describing himself

> as an evolutionary evangelist, he underlines the fervor of his cause and the inevitable collision between evolutionary theory and biblical Christianity. In sharing his sense that preaching the New Atheists as the prophets of God is his supreme calling, he points us to what is ultimately at stake.
>
> We are engaged in a great battle for ideas that Christians understand to be a battle for hearts, minds, and souls. Dowd and his fellow evangelists for evolution are certain that they own the future, and that biblical Christianity will simply fade and disappear. "Ours is a time of space telescopes, electron microscopes, supercomputers, and the worldwide web," he asserts. His conclusion: "This is not a time for parsing the lessons given to a few goatherds, tentmakers, and camel drivers."
>
> Well, give Michael Dowd credit for reminding us where the rejection of biblical Christianity inevitably leads.

Where does it lead? Away from herding goats, making tents, and driving camels, and toward space telescopes, electron microscopes, supercomputers, and the worldwide web? That's plainly what Dowd is telling us. I don't think that that is probably the message Mohler wants to give, though.

But I want to focus on one curious phrase: by leaving Christianity, Mohler says, we are facing a "theological and biblical cost." He believes we pay a price for throwing out faith and old religions, but he's vague on what it is, except that it's "biblical" and "theological."

Well. If the price of being an atheist is to carry a kicking, squalling Theology out to my backyard altar, stick a knife through its rancid little heart, and burn its noisome fat and bones in an offering to Science, I'm willing. It's hardly necessary; I think we can just watch it fade away into irrelevancy, like the long-faded scholarly traditions of divining the future from the guts of slaughtered animals or from the flight of birds. Telling an atheist that his or her actions will lead to the loss of schools of theology is no deterrent at all—Mohler might as well tell me that practicing regular bathing will cause me to lose the comforting company of lice. I don't think I'm unusual, either: if Mohler were to query your average Baptist-on-the-street, most wouldn't give a good spit for theology; they're believers in tradition, authority, and social conventions, not in the twisty weird rationalizations of seminarians. *Theology* is a magic word to reassure them that some smarty-pants somewhere has a reason for the rituals they go through, but it's hardly necessary.

What about this mysterious biblical price? If it is giving up the Bible wholesale, no—it's freely accessible online. I'd even say that it's an indispensible work of literature in Western history, so it should continue to be taught and considered, just as we do the Gilgamesh epic and the works of Thucydides and Herodotus. So the price is not even so much as the loss of a book.

Perhaps Mohler is referring to the promises of heaven and hell in his holy book? Promises are cheap, and unlikely promises backed up by no collateral and no certifiable authority are worthless. Empty threats are just as useless: Mohler could announce that he had been planning to leave me a million dollars in his will but then decided not to reward my heathenish ways. I'm sorry, but I would not pine for my lost fortune at all. It never existed. An imaginary god can rage all he wants in the pages of the Bible, but I lose nothing by ignoring him.

I wonder if some Christians resent atheists because Christians pay a burdensome price—they sit through tedious church services, make offerings, waste time, endure the absurdity of the whole phenomenon—while atheists escape all that with no apparent loss of anything. Oh, but atheists will pay eventually, they reassure themselves, either with an eternity in hell (while they smugly look on) or now, in the real world, with the loss of theology and the Bible. Oh, no! I shudder in fear!

But believers don't know the price they pay, the loss they suffer right now. I don't mean the trivial loss of sleeping in on Sunday morning, something I don't do myself, but the loss of a human perspective in favor of misplaced priorities that screw up their own lives and the lives of those around them. People matter far more than theology, and crazy dogmatists like Mohler have lost sight of that. I told you he was hilarious but also a bit scary; for an example of the latter, look again at the first sentence I quoted above.

> I regret to learn of Michael Dowd's cancer, but my concern for him is far more urgently focused on his malignant beliefs.

Wow.

Dowd has cancer, a disease that could cause him pain, suffering, and death, but Mohler considers that a matter of lesser concern than the fact that he doesn't accept Baptist doctrine. If any of you have heard Dowd, you know that he's enthusiastically cheerful and seems to be coping well—

his "malignant beliefs" don't appear to be interfering with his rational and effective response to illness. What cost is he paying?

Maybe I'm odd, but if I were arguing with someone who was ill, whether it be Michael Dowd or Albert Mohler, I would never consider it far more important for him to convert to atheism right now than to take care of himself with the best of modern medicine. Not that we couldn't continue to argue, but have some perspective—reality should always be priority number one.

Maybe the price atheists pay is the loss of childish delusions and the surrender of the idea that ghosts and fairy tales are more important than human lives. And yet we don't even have to pay garbagemen to haul that junk away! Perhaps the sacrilege here is that we get such an unbelievably good deal, and they don't.

Imagine No Heaven

Religion has a real problem with incentives. As long as the goodies are all in an invisible afterlife, it's hard to take them seriously.

The religion I grew up in was rather vague about the consequences of good and bad behavior. There was a hell, which was not discussed in polite company, and heaven was a place brought up at funerals in answer to the question of where the dead had gone. I didn't think much about it until I was in my early teens, when a crazy lady forced me to—and mainly, she made me realize she was a crazy lady.

One day my brother and I were walking down James Hill in our hometown of Kent, Washington, when we saw a woman yelling and hallooing at us, waving her arms frantically. We stopped—we were young and trusting and living in a small town. Nowadays I'd probably start running, pull out a cell phone, and call the police. But like good polite boys, we asked what we could do to help her.

She waved a Bible and told us how happy she was to see us and to be able to bring us the Good Word. Had we been saved? she asked. And did we love Jesus?

She had a kind of manic cackle and a demented grin on her face, and despite our innocent Tom Sawyer–esque upbringing, I wondered if she was going to draw an obsidian knife and sacrifice us, laughing and smiling the whole time.

She really warmed to her subject when she began to talk about the Lake of Fire. "When you go to hell, you are thrown in the Lake of Fire"—she gave a big smile, and I noted she had very nice dentures—"and your skin will be burned off, and your lungs will be seared by the fumes. Do you know what brimstone is?"—smile—"It's a sulfurous rock so hot it melts and smokes, and it will be poured down your throat"—rapturous smile—"and you will be trapped there forever writhing in agony, and you can never get out, no matter how much you scream and scream and scream, and your mommy and daddy won't be able to save you, ever"—sad smile—"but I can save you now! All you have to do is swear your love for your Lord and Savior Jesus Christ, amen, right here, and you'll be spared the torments of the damned! Hee-hee!"

And then she told us to kneel on the gravel by the side of the road and put our hands on her Bible, which we did, because at this point I was afraid that if I didn't, our mommy and daddy would find our little corpses with our throats slit and a madwoman dancing in our blood. Then she recited some lengthy vow with lots of Jesus in it and looked at us expectantly with another mad-eyed grin. We mumble-whispered, "Yes, ma'am." She let us go, throats uncut, hearts still in our chests, heads still attached to our necks, and capered off triumphantly, having secured two more souls for her lord and master. Or so she thought.

But all she actually had done was to make me aware that people who believe in heaven and hell are freakin' nutbag insane.

This is a real problem for religion. Oh, sure, it works just fine as a motivating mechanism when it's a simply presented choice between pain and bliss (hint: most people choose bliss without even breaking a sweat). But if you really think about it, the whole concept of heaven and hell doesn't hold up under inspection at all, and the metaphysical carrot and stick motivates behavior only if nobody tries to think it through. It's not just that there's zero evidence for an afterlife, but claims that it exists are logically untenable.

Hell. Let's start with the ol' bugaboo, a bad place of eternal suffering. Most of the naïve portrayals of hell are weirdly material. But if I'm dead, I'm bodiless, so what's all this nonsense about pain and body parts on fire? And really, I don't think my mind, or anyone's mind, could hold up for more than a few moments under the kind of intense agony these depictions propose before consciousness is lost, insanity ensues, and a complete disintegration of self and personality is accomplished. Simmering a twitching husk in a pool of lava for eternity seems rather pointless. Unless, of course, God is some kind of mindless psychopath who likes pointless torture.

Other visions of hell are a bit more sophisticated: it's a place of psychological torture, unending despair, and futility, where you feel regret and sorrow for all time, or suffer because you are deprived of the presence of God. That's a bit more plausible for a disembodied self, I suppose, but still—throw a mob of people into a Slough of Despond for a long, long time, and at some point people are going to get together and form a glee club, and so there will be singing in hell. And then a rugby match will break out, and there will be cheering and betting, and thespians will be pestering Shakespeare to write some new plays, and before you know it, culture will emerge, and it won't be hell so much anymore. Dante's glum ghosts never seemed particularly likely to me.

But all right, let's assume God has figured out ways to permanently suppress the human spirit among all the deceased and has contrived a truly painful hell, one that I cannot imagine but that he can, being God and all. Now we've got the problem that the loving God we're all supposed to worship is an imaginative, creative death camp commandant who also maintains a luxury spa on the side. In that case, let's just scratch *loving* and *worship* from his description.

Here's another twist. Most of the people who believe in this pain-god also believe in a final judgment—that up to a certain point, you've got an opportunity to make the right choices and get in God's good graces, and after that—sorry, you're just done. You've abandoned all hope of salvation and will suffer forever, with no possibility of forgiveness. Does that make sense? No. I can understand fear of punishment as a motivator—maybe angels whose harps fall out of tune get slapped with an hour in the Lake of Fire so they don't do it again—but what's the point of punishment without reprieve, without opportunity for reform? Your soul gets the first brief seventy years of its existence to decide whether Mormonism or Islam or Lutheranism (ELCA) or Lutheranism (Wisconsin Synod) or Lutheranism (Missouri Synod) or Catholicism is true, and if you make the wrong guess, you have to spend the next billions and billions and billions of years in utter abject misery? That's insane.

The whole concept of hell is so demented and irrational that many religious people have abandoned it. Quite a few Christians sensibly reject it and simply say that because their god is a loving god, everyone gets to go to heaven. Of course, now they're stuck with the concept of heaven.

Heaven. The funny thing is, for all my inability to imagine a viable vision of hell, what probably comes closest is most people's version of heaven. Who would want to even visit Puritan Heaven, with all its smug and judgmental inhabitants praising God nonstop with pursed lips and suspicious eyes? What woman would want to live in Mormon Heaven, and what man could dwell there for long without developing a smidgen of guilt?

Heaven has all the problems of hell. Just as my personality, my self, me, would not persist if I were immersed in the Lake of Fire, making immortality rather irrelevant, so too would I cease to exist if I were scoured in the all-consuming love of an omnipresent deity. What kind of PZ Myers could exist when stripped of doubt and disbelief and irreverence and impudence? An angelic PZ is a contradiction in terms—it would be the death of the me that counts. I am disbarred by my nature from any form of heaven.

A paradise would be inhuman. (I know, one can get around this problem by arguing that after death you aren't human anymore by definition; but that requires throwing away the idea of life after death, which is what most people find appealing.) Think about what defines you now: how you think, your personality, your desires and how you achieve them—what you strive for. Finish one project, and what do you do (after a little celebration, of course)? You look for something else to strive for, a new goal to keep you interested and occupied. But now you're in heaven. All your wishes are fulfilled, all your desires achieved, you're finished with everything you've ever dreamed of. That makes heaven a kind of retirement home where everyone is waiting to die. Waiting forever.

You can't ask human beings to shed all their aspirations without expecting them to seek new ones, and no, membership in the Choir Celestial isn't going to be everyone's aim; nor is doing the finest job in the universe buffing God's toenails. Given eternity to plan, you just know everyone is eventually going to get around to plotting the Big Coup, and then it's back to hell again, which we've already said is an untenable concept.

Another version of heaven rips away all those troubling desires that define our humanity. This heaven is pure bliss, pure joy, pure worshipful existence in the presence of the deity; you won't miss those trifling hopes because they will be blown away by unadulterated rapturous ecstasy fired straight into your soul by Jesus himself. The high will never ever end.

This is the crack cocaine vision of afterlife.

I confess, it does have some appeal.

It is, once again, a kind of death of the self, an end of your existence as you know it in favor of mindless nonexistence, but at least it's winking out cheerfully. It's rather like being given a choice on your deathbed: you can (1) have an overdose of painkillers with a tincture of ecstasy and aphrodisiacs while having aorta-rupturing sex; or (2) go out gasping and choking and clawing for a few more minutes of strained life, fighting for every second. I'd tend to favor number two myself, but I can sympathize with those who would rather pick choice number one. Either way, though, it's an end of who you are.

Some religions embrace this sublime vision of an ultimate end that doesn't include the mundane humanity of its believers. Buddhism's afterlife, for one, does seem to be a kind of selfless oblivion. But the Abrahamic religions have still got the cartoonish anthropocentric version of afterlife, where you've got a body with limbs and a tongue and a penis or vagina, and you get to indulge in the senses within certain confining rules. You get to meet Grandma and Grandpa again, and they aren't all subsumed into the godhead—they're there to give you hugs and a plate of cookies. And that's just silly. I can't believe a word of it.

I especially cannot believe any of it in the absence of reasonable evidence. It's really nothing but people making stuff up based entirely on what they wish were true.

I'm not a popular fellow at funerals, as you might guess. It's not that I'm a disruptive shouter—funerals are places for grief and consolation, not anger—but I'm not going to deliver a eulogy unless it's something like "The deceased is dead and gone forever. Now get on with your lives and accomplish as much as she did. Oh, and by the way, that smug jerk in the funny collar over there is lying to you if he says anything different."

The godless view of death is that it's not a transition—it's an end. It deserves all the sorrow that the living bring to it, but the absurd attempts of believers to soften it with lies are a contemptible disservice to the life that is over. Don't be shy about saying it; the life that is lost is far more important, and the death far more tragic, than any fantastic superstition conjured up by some pious twit.

Come to think of it, though, funerals are fitting places for sacrilege. You're all invited to mine, someday, and please do have a laugh over that rotting meat at the front of the room. I'll be gone, I won't mind—but please be nice to the family, even if they're right up there razzing God with you.

Daughters of Eve

Half the people in the world commit this sin against God: they are born female.

It's astounding that any women at all accept Christianity, Judaism, or Islam; these are profoundly misogynistic faiths. Throughout the Christian Bible, women are treated as chattel to be abused and misused, and uppity women are regarded as the worst of the lot, fit only to be slaughtered. Parts of the Bible read like snuff porn—but it's all okay, because it's the Bible, God's holy word, and if God is gonna have to choke a bitch, who are we to question it?

We can trace the attitude right back to Genesis 3:16. The Bible begins with a story that blames women for all the pain of sin in the world, so women need to be put in their place, under the dominion of men.

> Unto the woman he said, I will greatly multiply thy sorrow and thy conception; in sorrow thou shalt bring forth children; and thy desire shall be to thy husband, and he shall rule over thee.

When I started looking up woman-hating Bible quotes, I was a bit overwhelmed; they're everywhere. The many contemptuous words about women that you can find there would flood this essay. I won't do that, but I will give a few examples. Job 14 is an interesting one, because it at least acknowledges that men are damned, too. But why are they so troubled? Because they're born of those filthy, dirty, nasty women.

> Man that is born of a woman is of few days and full of trouble. He cometh forth like a flower, and is cut down: he fleeth also as a shadow, and continueth not. And doth thou open thine eyes upon such an one, and bringest me into judgment with thee? Who can bring a clean thing out of an unclean? not one.

I don't like how the Bible talks about my mother.

This wasn't just an Old Testament aberration, superseded by the loving Jesus of the New. The apostle Paul was just as terrible a misogynist. Here's the classic 1 Timothy 2:8–15:

> I will therefore that men pray every where, lifting up holy hands, without wrath and doubting. In like manner also, that women adorn themselves in modest apparel, with shamefacedness and sobriety; not with broided hair, or gold, or pearls, or costly array; But (which becometh women professing godliness) with good works. Let the woman learn in silence with all subjection. But I suffer not a woman to teach, nor to usurp authority over the man, but to be in silence. For

> Adam was first formed, then Eve. And Adam was not deceived, but the woman being deceived was in the transgression. Notwithstanding she shall be saved in childbearing, if they continue in faith and charity and holiness with sobriety.

Shut up, ladies—it's your fault we got kicked out of the Garden of Eden, and you're all going to have to pay for it by suffering in silence.

I know the next objection apologists will make: we shouldn't read the Bible so literally, and you atheists are just as bad as the fundamentalists. The Bible has to be carefully interpreted, and we need wise men (of course) to extract the deeper truths. So let's go along with that and ask a few of the church fathers what it all means.

Let's see what Tertullian has to say—he's always fun.

> Woman is a temple built over a sewer, the gateway to the devil. Woman, you are the devil's doorway. You should always go in mourning and in rags.

Whoa. Tertullian is one of those fellows you'd really like to see psychoanalyzed; he's one warped little psycho thug. Maybe he's not representative at all, and perhaps we should look to a few Catholic saints, who will certainly be more enlightened.

Here's Saint John Chrysostom:

> Among all savage beasts, none is found so harmful as woman.

Ooops. Maybe Saint Clement will be friendlier.

> Every woman should be overwhelmed with shame at the thought that she is a woman.

Saint Augustine?

> Any woman who acts in such a way that she cannot give birth to as many children as she is capable of, makes herself guilty of that many murders.

Now that's a Catholic attitude. Perhaps Catholicism is the problem here, and we really need to look to the Protestant theologians. Here's what Calvin had to say:

> Woman is more guilty than man, because she was seduced by Satan, and so diverted her husband from obedience to God that she was an instrument of death leading to all perdition. It is necessary that woman recognize this, and that she learn to what she is subjected; and not only against her husband. This is reason enough why today she is placed below and that she bears within her ignominy and shame.

Same old story. I was brought up Lutheran, which meant that all I ever heard about that deranged founding cleric was stories about his iconoclasm and bravery. It took me a few years, after leaving the church, to discover that he was also utterly insane about witches and demons, and that he was an anti-Semitic monster who set the bar high for the Nazis. But he also praised marriage and thought priestly celibacy was unnatural and wrong, so could he have been a bit less sexist than others? Alas, more disappointment.

> If a woman grows weary and at last dies from child-bearing, it matters not. Let her die from bearing, she is there to do it.
>
> Women have narrow shoulders and wide hips, therefore they ought to be domestic; their very physique is a sign from their Creator that he intended them to limit their activity to the home.

So Christianity regards women as being of relatively little worth. What is to be done with them, then? And here's where the Bible begins to read like a script for a slasher horror movie (which also, by the way, are as vicious to women as anything in the Bible—so this is also a secular issue). One verse that has cost thousands of women their lives, and that is still being used to justify murder, especially in Africa, is Exodus 22:18.

> Thou shalt not suffer a witch to live.

Got women usurping the role of priests? Kill them. If they're just standing around innocently, then read Deuteronomy 20 and Numbers 31 for the appropriate response: rape them.

> But the women, and the little ones, and the cattle, and all that is in the city, even all the spoil thereof, shalt thou take unto thyself; and thou shalt eat the spoil of thine enemies, which the LORD thy God hath given thee.

> Now therefore kill every male among the little ones, and kill every woman that hath known man by lying

> with him. But all the women children, that have not known a man by lying with him, keep alive for yourselves.

And of course, you can be brutal in your destruction of women. Second Kings 9 tells the story of a war between two factions, the side of King Ahab and Queen Jezebel, who promoted the worship of Baal, and the faction of the prophet Elijah, who favored the Hebrew God. Here's what happened when King Jehu and Elijah finally defeated the Baal worshippers.

> And when Jehu was come to Jezreel, Jezebel heard of it; and she painted her face, and tired her head, and looked out at a window. And as Jehu entered in at the gate, she said, Had Zimri peace, who slew his master?
>
> And he lifted up his face to the window, and said, Who is on my side? who? And there looked out to him two or three eunuchs. And he said, Throw her down. So they threw her down: and some of her blood was sprinkled on the wall, and on the horses: and he trode her under foot.
>
> And when he was come in, he did eat and drink, and said, Go, see now this cursed woman, and bury her: for she is a king's daughter. And they went to bury her: but they found no more of her than the skull, and the feet, and the palms of her hands. Wherefore they came again, and told him. And he said, This is the word of the LORD, which he spake by his servant Elijah the Tishbite, saying, In the portion of Jezreel shall dogs eat the flesh of Jezebel: And the carcase of Jezebel shall be as dung upon the face of the field in

> the portion of Jezreel; so that they shall not say, This is Jezebel.

See? Snuff porn!

The other weird twist to this story is that it is about a war. Jezebel was a fierce queen who fought for her side and actually had some significant political clout. Somehow, though, her name has become transformed to mean a harlot, a vain and promiscuous woman—all because in that first verse above, she bothered to put on her makeup before the conquering army of Elijah slaughtered her. Obviously, then, she deserved to be tossed out a window, trampled by horses, and fed to the dogs.

Now I have to confess, I don't think religion is the root cause of misogyny. Religion is to misogyny as disease is to misery—not the sole cause, but a significant contributor. I'm going to blame biology.

We are a somewhat sexually dimorphic species. Males do tend to be, on average, larger, more muscular, and more aggressive than females, and that means that in our crude history one strategy for reproductive success has been brute force: men controlling women to compel them to bear children for them. The Bible wasn't the cause of all those men raping women, because conquering men were doing that long before Jehovah—it was merely endorsing the status quo.

But here's the thing: Sometimes we know we can do better than what our brute natures tell us to do, and by transcending our instincts, by using our brains and our capacity for reason, we can become greater than before. Any culture that denigrates half its members, that commits them to inferior roles by tradition rather than allows them to excel by inclination, is throwing away half its potential. People do more than make babies, and if you value people doing science, writing poetry, creating bridges and buildings, and designing software, you can't rely on the Stone Age virtues of bulk and testosterone. Our dimorphism isn't relevant—it's brains that matter.

What religion has always done is pander to biases, achieving success not by challenging people with something new but by reinforcing what they already believe, no matter how wrong it is. By contrast, science shines by breaking the shackles of our preconceptions and letting us work with what really is.

The hard question, though, is why women ever fell for women-hating religions in the first place—and women have undeniably been prominent advocates of the church, of submission, of labeling themselves in dress and manner as toys of men. I have my own explanation, which is that religion is a kind of parasite of the mind that promotes its own disease. Where does religion have its greatest success? Among the miserable and oppressed, because it is very good at promising (but not delivering) hope. If you are among the downtrodden, magical answers have a great appeal: pray and you shall be delivered, believe and you shall be exalted in heaven. These answers do nothing but make the believer feel better about his or her problems. In fact, they actually increase the misery by encouraging believers to shun productive solutions in favor of nonanswers. Misery leads people to turn to religion, which can make the misery worse, because religion doesn't address the material causes of the misery, and the increasing despair leads to more and more seeking succor from religion, which makes it even worse, which . . . You get the idea. It's a spiral of futility.

And who has been slapped down most consistently throughout human history? Women. It's not surprising that the segment of society that is oppressed is also one that often turns to faith to gain the illusion of relief.

But let's close with a ray of real hope, and surprisingly, it comes from the contempt of Tertullian.

> Each of you women is an Eve . . . You are the gate of Hell, you are the temptress of the forbidden tree; you are the first deserter of the divine law.

Tertullian thought he was saying something bad there, but think about it. Every woman is an Eve, and Eve is the one who led humanity to the knowledge of good and evil. She was the "first deserter of the divine law." Eve was the first freethinker! When you turn it around that way, it's actually a very nice myth. There is part of the answer to misogyny: break free of the old dogmas that want women to be submissive bearers of babies and nothing more, break the dead and oppressive divine laws, and tempt men to follow you into liberation. We'll all be free together.

Take pride in the example of Eve—she is the author of a real promise of a greater humanity.

Prometheus's Sin

Why are science and religion in conflict? Because changing ideas and new knowledge are sacrilegious.

Throughout Ken Ham's Creation Museum, in northern Kentucky, a persistent story is exhibited in display after display. Two ways of looking at the world are shown: "God's Word," the ultimate source of knowledge, the Bible; and "Human Reason." For Christians, human reason is always the fall guy, the error-filled path, while the only truth lies in listening to what God has to say. Christians have an old book with the whole story laid out—literally, as the creationists like to claim—and by their definition, all observations about the natural world must be accommodated to it. In contrast stands human reason, which dares to contradict the Bible, dares to show great truths not encompassed by the Bible's stories, and most horribly, proposes an alternate, better source of knowledge than a body of ancient myths.

That's a major theme throughout the "museum," that science defies the word of God, and that the only valid knowledge must be that which is reconcilable with the Bible; Scripture is the sole arbiter of truth.

According to the Creation Museum,

> In a biblical worldview, scientific observations are interpreted in light of the truth that is found in the Bible. If conclusions contradict the truth revealed in Scripture, the conclusions are rejected.

To that mind-set, insisting on the primacy of evidence other than the Bible is heretical—a theme at the evangelical Christian creationist organization Answers in Genesis, for instance, is that even the phrase *millions of years* is a signifier of gross, un-Christian error, since the Bible clearly (doesn't it?) explains that the earth is only six thousand years old.

But, you might say, isn't fundamentalist Christianity a kind of pathological religion that carries its antirational claims to absurd extremes? Is it fair to judge faith in general on the basis of this one radical example? Yes. Because fundamentalist Christianity isn't at all unusual. Consider that well-known sixteenth-century theologian Martin Luther. Oh, Luther offers a rich vein of distressing statements opposing rationality.

> Reason is the Devil's greatest whore; by nature and manner of being she is a noxious whore; she is a prostitute, the Devil's appointed whore; whore eaten by scab and leprosy who ought to be trodden under foot and destroyed, she and her wisdom. . . . Throw dung in her face to make her ugly. She is and she ought to be drowned in baptism. . . . She would deserve, the wretch, to be banished to the filthiest place in the house, to the closets. Reason is the greatest enemy that faith has; it never comes to the aid of spiritual things, but—more frequently than not—struggles against the

> divine Word, treating with contempt all that emanates from God.
>
> People gave ear to an upstart astrologer [Copernicus] who strove to show that the earth revolves, not the heavens or the firmament, the sun and the moon. Whoever wishes to appear clever must devise some new system, which of all systems is of course the very best. This fool wishes to reverse the entire science of astronomy; but sacred scripture tells us [Joshua 10:13] that Joshua commanded the sun to stand still, and not the earth.

Note that last objection: this is not just the opinion of some radical Protestant. The idea was shared with the Catholic Church, which similarly resisted the conclusions of astronomers. Islam also promoted geocentrism, despite the fact that the Koran is said to be without error and contradiction. That's the problem with having a source that is claimed to be infallible but was actually written by people who knew next to nothing about the world around them—the stories don't hold up.

Unfortunately, the religious strategy for coping with this conflict is not to maintain flexibility and adapt to new information, but instead to restrict new knowledge and condemn it when it contradicts tradition.

At the very least, religion's fear of honest information about the world leads to stagnation; at worst, it is destructive to any culture that values scientific advances and the education of its children. Here's a nightmare to contemplate: the staff of Answers in Genesis teaching children about science. And they do! They lead groups of children through recitations condemning evolution and all science that denies the "facts" of the Bible, sing songs about how the earth is only six thousand years old and the dinosaurs sailed on the Ark with Noah, and teach them how to stump scientists. (It's easy: ask scientists "Were you there?" and when they say no, you've demonstrated that they have no evidence to back up their science.)

I'm beginning to think that child abuse is a tenet of the Abrahamic religions.

So here are some more sacrilegious acts you can commit: Learn something new. Teach something new. Question dogma. Challenge tradition. Laugh at the quaint myths religion offers us.

So Alone

Scientists and atheists do something that many believers find repellent: we shatter their perception of their relationship to the universe. And understandably, they don't like that.

Most religious people in the West have a specific model of how the world works that is based on our cultural history as the progeny of nomadic herders and that still resonates strongly with all of us—the father-child relationship, the patriarchy. We have a wise leader who guides us all, punishes us when we stray, offers largesse to those in his favor, and unites the whole tribe in common cause. Those Bronze Age sheepherders lived this way, and for them patriarchy made sense. It was a strategy for survival that worked well. But it shapes the way we see the world even now. Ask any Christian on any Sunday morning about flocks and sheep and shepherds, and he or she will understand the metaphor, despite never having been in contact with animals other than a household pet.

Patriarchy is also a powerful idea because it posits a set of very personal relationships. The father is remote because of his great responsibilities, but at the same time, we all want that pat on the shoulder, the encouraging word, the opportunity to serve and win distinction in Father's eyes by virtue of our dutifulness. It's a familial relationship, tightly knit and long-established, wherein we are respectful dependents and the leader of the tribe relies on our service.

Beyond the family and tribe, though, this vision has been extended to the entire universe. A great Patriarch in the Sky is our leader and guide, responsible for making the grand strategic decisions about where our tribe will go, and watchfully making sure that the tribe's unity is not disrupted by wayward ideas from nonconformists. His central concern, which we all share, is that our people should thrive, and even if he is stern at times, it is because he cares so much that we succeed. And he knows each one of us personally, just as the leader of tribe or clan in pastoral days would have, and he can give us an approving stroke or a damning, angry smiting, depending on whether we help or hinder the work of getting the flocks to the summer pasturage.

Read your Bible. It's saturated with this primitive herdsman mentality: God the father, the sheep and goats, the lost lambs and the Lamb of God, flocks and herds. It's anthropologically fascinating and not necessarily an evil metaphor (unless you're a woman—the patriarchy is deeply misogynistic). One of its most appealing aspects is that it makes one's relationship with the universe close and personal, in a very simple kind of relatedness, that of father and child. It reduces all the complexity of deep time to the metaphor of a single generation, direct and immediate, and it colors everything about how we view our place in the world: domination and submission, leader and follower, wisdom and naïveté, master and servant, command and obedience. Biblical morality also tangles up our material relationship with the world in those paternal virtues of love and concern and discipline. Its patriarchal emphasis muddles ethical concerns, which fits with the portrait of the omnipotent God painted by the Bible.

I can sympathize. I loved and respected my father, and any attempt by an outsider to defame or complicate or diminish him would trigger a resentful response from me. Christians and Muslims and Jews have been told from their earliest years that God is their father, with all the attendant associations of that argument, and what are we atheists doing? Telling them that, no, he is not, and not only that, you don't have a heavenly father at all—the imaginary guy you are worshipping is actually a hateful monster and an example of a bad and tyrannical father. And you aren't even a very special child—you're a mediocre product of a wasteful and entirely impersonal process.

By comparison, that whole business of breaking the news about Santa Claus is small potatoes. Reality is harsh, man.

But it is reality. We've done the paternity tests, we've traced back the genealogy, and we're doing all kinds of in-depth testing of the human species. People are apes and the descendants of apes, who were the descendants of rat-like primates, who were the children of reptiles, who were the spawn of amphibians, who were the terrestrial progeny of fish, who came from worms, who were assembled from single-celled microorganisms, who were the products of chemistry. Your daddy was a film of chemical slime on a Hadean rock, and he didn't care about you—he was only obeying the laws of thermodynamics.

You are you not because of some grand design but because of chance, contingency, and selection. Your genome is a mess of detritus with a tiny fraction of well-honed functionality, and your body is cobbled together from the framework of a tetrapod. You bear the scars of chance throughout, and you are mostly unaware of them because selection—that is, the death of millions—has patched them over. But they're present to the eye that will look for them. You aren't the best at much of anything: you're weaker, slower, more fragile, and clumsier than the other species we compete with, and although you've got a bigger brain, the majority of Americans, at least, consider it a virtue to keep it ignorant and unused. Universally, we have difficulty thinking in the long term, while we are very good at exploiting our environment in the short term, which is leading to some interesting and possibly fatal consequences.

The legacy of good husbandry, we are not. Our cosmic father did not and does not exist, which is a good thing, because if he did, he's the kind of lazy, destructive deadbeat we'd be ashamed of.

This is our new heresy. We have killed our heavenly father, demolished that cozy personal (but imaginary!) relationship with a great and caring being. We are alone, orphans in an indifferent universe. We atheists must be a cold and broken people, without hope, without love.

But of course, we're not, and I think this change in our vision of our relationship to the universe is humankind's great good hope. Primitive monotheistic religions have shackled us to a limited metaphor and model, the father and child, and erected an entire invisible heavenly mouthpiece to help us maintain that comfortable delusion—but it's like relying on the Great and Powerful Oz to help us out of our problems, when Oz is only a sham and a show. We have to escape this narrow perspective.

Reality doesn't just destroy the patriarchal model—it

gives us new and better ways to visualize our relationship with the universe. Father-and-child is inadequate; we have to think in terms of populations and species interacting (not dominating), of being part of an environment. There is more to life than the father-child bond. I am the outcome of a trillion coalescing possibilities, with a vast population of brothers and sisters acting out our brief lives on a background of gas and stone, water and light, grasses and fishes, and my responsibilities are far greater than obedience to a father figure. Breaking that illusion of a personal tie to one grand elder lord can briefly leave us feeling abandoned and alone and lost, and I can understand how some people find severing that imaginary relationship a horrible prospect.

But here's the wonderful revelation. If you're a well-adjusted person, once you've discarded the unhealthy fictitious relationship with a phantasm, you can look around and notice all those other people who are likewise alone, and you'll realize that we're all alone together. And that means you aren't alone at all—you're among friends. That's the next step in human progress—getting away from the notion of minions living under a trail boss, and moving onward to working as a cooperative community, with no gods and no masters, only autonomous agents free to think and act.

One Nation Free of Gods

A strange thing has happened to this country: somehow the United States of America has become a biblical entity. I know, the country didn't even exist for way over a thousand years after the Bible was composed and assembled, and there isn't one word about the USA in the text, but you couldn't tell from the way some people have confused patriotism and piety.

In 1935 Sinclair Lewis wrote a novel called *It Can't Happen Here* about an America taken over by a populist dictator. His hero explained how that could happen:

> "Why, there's no country in the world that can get more hysterical—yes, or more obsequious!—than America. Look how Huey Long became absolute monarch over Louisiana, and how the Right Honorable Mr. Senator Berzelius Windrip owns HIS State. Listen to Bishop Prang and Father Coughlin on the radio—divine oracles, to millions. Remember how casually most Americans have accepted Tammany grafting and Chicago gangs and the crookedness of

so many of President Harding's appointees? Could Hitler's bunch, or Windrip's, be worse? Remember the Ku Klux Klan? Remember our war hysteria, when we called sauerkraut 'Liberty cabbage' and somebody actually proposed calling German measles 'Liberty measles'? And wartime censorship of honest papers? Bad as Russia! Remember our kissing the—well, the feet of Billy Sunday, the million-dollar evangelist, and of Aimée McPherson, who swam from the Pacific Ocean clear into the Arizona desert and got away with it? Remember Voliva and Mother Eddy? . . . Remember our Red scares and our Catholic scares, when all well-informed people knew that the O.G.P.U. were hiding out in Oskaloosa, and the Republicans campaigning against Al Smith told the Carolina mountaineers that if Al won the Pope would illegitimatize their children? Remember Tom Heflin and Tom Dixon? Remember when the hick legislators in certain states, in obedience to William Jennings Bryan, who learned his biology from his pious old grandma, set up shop as scientific experts and made the whole world laugh itself sick by forbidding the teaching of evolution? . . . Remember the Kentucky night-riders? Remember how trainloads of people have gone to enjoy lynchings? Not happen here? Prohibition—shooting down people just because they MIGHT be transporting liquor—no, that couldn't happen in AMERICA! Why, where in all history has there ever been a people so ripe for a dictatorship as ours! We're ready to start on a Children's Crusade—only of adults—right now, and the Right Reverend Abbots Windrip and Prang are all ready to lead it!"

Change a few names and that could have been written today: all it needs is Billy Graham, Pat Robertson, "freedom fries," the Texas Board of Education, Rush Limbaugh, and Guantánamo Bay tossed in to be brought up to date. Nothing has changed; if anything, we've gotten worse over the years.

The book can be summarized in this misattributed quote (Lewis didn't actually say it, but it is a perfect description of our situation):

> When fascism comes to America, it will be wrapped in the flag and carrying a cross.

Look at America today and we see this phenomenon in progress: think tanks, ideologues, and religious fanatics insist that we are a Christian nation—and when we are most afraid of external threats, what do they do? Entangle the country deeper in the web of the sacred. We saw it at the height of the Cold War, when the Communist threat was used to excuse violations of the separation of church and state. A National Day of Prayer was declared in 1952; the president was roped into a yearly National Prayer Breakfast in 1953; the phrase *under God* was added to the pledge of allegiance in 1954; "In God We Trust" was decreed the national motto in 1956. Coincidence? Of course not. Patriots saw an opportunity to make nationalism sacred, and religious fanatics saw a chance to tie support for their dogma to the state.

It's happening now, too.

All you have to do is look to the Texas Board of Education and its recent overt embrace of Holy America. It is rewriting history to create an America born out of the overt action of divine providence. Cynthia Dunbar, a board member, wrote:

> Whether we look to the first charter of Virginia, or the charter of New England or the Charter of Massachusetts Bay, or the Fundamental Orders of Connecticut, the same objective is present: a Christian land governed by Christian principles. I believe the entire Bill of Rights came into being because of the knowledge our forefathers had of the Bible and their belief in it. . . . I like to believe we are living today in the spirit of the Christian religion.

Perfect. Do not question God. Do not question America. Mix those two ideas together, and you've got a lovely recipe for blind obedience.

I usually complain about religion, but I have to add another target: patriotism. It's the same thing, opening a door to unthinking authoritarianism, and it always leads to oppression. Quite contrary to the claims of fanatical Christians, the heart of a thriving democracy has to be constant questioning of the operation of the government. To marry religion to our government would be antithetical to its founding principles, and even to regard those founding principles as inviolate and somehow imbued with godly authority would be a betrayal.

When I was growing up in the 1960s and 1970s, defenders of the status quo threw around a common slogan: "My country, right or wrong." Even at my young age, that always seemed insane. If my country is wrong, shouldn't I want to change it?

Question religion. But also question your government. It wasn't founded by gods.

An Embryo Is Not a Person

I have a frank admission to make: I am pro-abortion. I'm not of the weaselly ilk that professes to deplore abortion but supports tolerance because the alternatives are appalling. I say, give the blastocysts the heave-ho without reservation. Furthermore, I find the willingness of the devout anti-choice mob to distort the evidence so appalling that I would not join with them in *anything*.

When I gave a talk at the University of Wisconsin at Stevens Point in 2009, a few protesters showed up outside the building—friendly, nonviolent people. I invited them in to listen. Still, the group picketing me were Catholic right-to-life weirdos. I've often encountered the most angry opposition from critics who focus not on my atheism but on my disregard for embryos. This particular group in Wisconsin was handing out flyers complaining about my desecration of crackers.

> Myers hammered his rusty nail while gleefully boasting that "nothing is sacred." One has to wonder if he would likewise gleefully drive nails through the hands

and feet of Christians he so publicly despises?! Obviously Christians themselves are not sacred to him.

There they go again, making my point for me. I do not think Christians or crackers are sacred, not because I devalue human life, but because "sacred" is an invalid rationale for doing so; the value comes from the individuals themselves, not from some imaginary decree from a nonexistent ghostly entity. It is also a great shame that Catholics so obliviously and so willingly equate themselves with crackers; they assume that because I would abuse a piece of bread, I would treat human beings in exactly the same way—apparently they think that cracker is just as precious as a person. (Actually, it's much tastier.)

That's a rather gross and dangerous error. A cracker is a flat piece of ground-up vegetable matter, baked and processed, mostly inert, sold with the intent of being further broken down in someone's digestive system. Throwing it in the trash in no way implies that a complex and dynamic being can be similarly disposed of or casually destroyed and consumed. *Sane* people have an appropriate perspective on the relative importance of foodstuffs and human beings. Crazy people can't tell the difference.

And speaking of crazy people—a kind of insanity destroys the ability of normal people to distinguish between embryos and people and encourages them to lie to others about the status of the embryo, treating it as the moral equivalent of a child. It's almost the same problem as revering a cracker over a person. A better question to ask is whether they would kill someone to defend a Jesus wafer. (Some, at least, have told me that they would.) Would they be willing to throw away their purported reverence for the sanctity of life in order to kill someone to defend a fetus? (We definitely know that some will.)

Certain groups are actively blurring the line between embryos and human beings, and I consider them just as wicked as the howling haters lined up outside women's health clinics—they use mistruths to foment attacks on people in order to defend nonpeople. One of the biggest, noisiest, and most dishonest such groups is Prolife Across America, which puts up billboards all across the United States; we have a similar outfit here in my state, Minnesota Citizens Concerned for Life. I see their billboards all along the roadsides on my drive into Minneapolis.

These groups' work is easily recognizable. They have one theme: fetuses and babies are exactly the same. All their signs feature cute baby pictures coupled to factoids about development, and they thoroughly enrage me. They are basically generating false associations about development.

A standard example declares, "Face it: I had my Eyes, Ears & even my Tongue! 28 Days from Conception," illustrated with a picture of a roughly one-year-old baby.

Awww, cute plump smiling baby with eyes and ears and a tongue—and did you know he formed those in the *first month* of pregnancy? How can you support aborting adorable little children!

It's even sort of weakly, tenuously true. The embryo does have nonfunctional primordia of those organs at twenty-eight days, but it's not exactly cute—it looks more like a worm. And it's only about 4 millimeters (or about 0.2 inches) long. Its "face" is a couple of branchial arches. Assuming they want to be honest in their advertising, I've

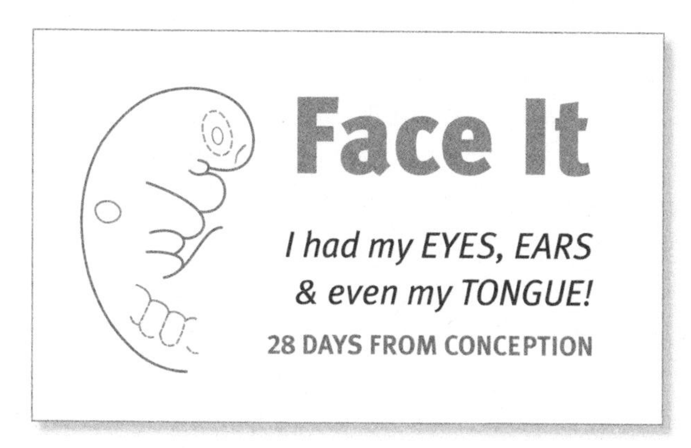

taken the liberty of redesigning their billboard for greater scientific accuracy. I suggest they add another line of copy, too: "I also had a tail!"

You get the idea. All of their billboards are like this, listing a known developmental landmark, the earlier the better, and showing a picture of a postnatal infant, as if that had any relevance. It's an attempt to make an emotional tie to developmental events. It works with a lot of people, too—people who are ignorant of reality. And, of course, these ads do nothing to educate them.

Another billboard states banal facts as revelations: "I got my genes at conception! I'm unique." It's accompanied by the ubiquitous stock photo of a smiling baby in overalls.

Yes, that baby did get its genes at conception. So? A collection of genes doesn't make a human being. There was no teeny-tiny infant spontaneously bursting into existence at the instant a sperm cell punched into an oocyte, immediately donning a stylish pair of OshKosh B'goshes.

This point is absolutely and solidly established in biology. The embryo is not the adult. It doesn't contain the full information present in the newborn—that will be generated progressively, by interactions with the environment and by complex internal negotiations within an increasingly complex embryo. Pretending that forty-six chromosomes in a cell are sufficient to define a person is the most absurdly extreme biological reductionism.

The fertilized oocyte is a human cell, but it is not a human being.

Way too many people think that is a sacrilegious idea—that we have to cherish every single scrap of human tissue, especially the bits that have the *potential* to go on and develop into a child.

No, we don't. We don't have to revere every block of rough marble because another Michelangelo could come along and sculpt it into something as wonderful as his *David*. We don't have to treasure every scrap of canvas because the next Picasso is going to use it for a masterpiece. The value is not in the raw materials but in the pattern, the skill, the *art* put into it. Similarly, those cells are simply the raw clay that the process and time will sculpt into something that is worth love and care.

Which is more important, the pigments or the painting? Even worse, do you think the pigments *are* the painting?

The Courtier's Reply

Criticisms of Richard Dawkins's *The God Delusion* share a refrain that is so common, to the point of near-unanimous universality, that I've taken to categorizing it with my own private title. I've decided to share it with you all, along with a little backstory that will help you to understand it.

I call it the Courtier's Reply. It refers to the aftermath of a fable.

> I have considered the impudent accusations of Mr. Dawkins with exasperation at his lack of serious scholarship. He has apparently not read the detailed discourses of Count Roderigo of Seville on the exquisite and exotic leathers of the Emperor's boots, nor does he give a moment's consideration to Bellini's masterwork, *On the Luminescence of the Emperor's Feathered Hat*. We have entire schools dedicated to writing learned treatises on the beauty of the Emperor's raiment, and every major newspaper runs a section dedicated to imperial fashion; Dawkins cavalierly dismisses them all. He even laughs at the highly popular and

most persuasive arguments of his fellow countryman, Lord D. T. Mawkscribbler, who famously pointed out that the Emperor would not wear common cotton, nor uncomfortable polyester, but must, I say must, wear undergarments of the finest silk.

Dawkins arrogantly ignores all these deep philosophical ponderings to crudely accuse the Emperor of nudity.

Personally, I suspect that perhaps the Emperor might not be fully clothed—how else to explain the apparent sloth of the staff at the palace laundry—but, well, everyone else does seem to go on about his clothes, and this Dawkins fellow is such a rude upstart who lacks the wit of my elegant circumlocutions that, while unable to deal with the substance of his accusations, I should at least chide him for his very bad form.

Until Dawkins has trained in the shops of Paris and Milan, until he has learned to tell the difference between a ruffled flounce and a puffy pantaloon, we should all pretend he has not spoken out against the Emperor's taste. His training in biology may give him the ability to recognize dangling genitalia when he sees it, but it has not taught him the proper appreciation of Imaginary Fabrics.

You would be amazed at how many of the anti-Dawkins, anti-atheist arguments can be filed away under the category of the Courtier's Reply. They rebuke us for ignoring "sophisticated" theology and the centuries of scholarship behind it. But our point is not to disparage study or knowledge. Rather, it's to point out that the scholars have not demonstrated the fundamental phenomenon underlying all that theological discourse—nor even consistently defined it. True, anyone can drop a stack of volumes in front of me that describe the nature of God, the desires of God, the history of God. But not one of those volumes can explain how they know God exists or how they determined the truth of all these elaborate fairy tales.

The Big Pink Guy in the Sky

Allow me to take a moment to describe the unsophisticated theology—the religious stuff that your average voter, politician, television pundit, and middle-class pulpit-pounding minister believes. It is the common faith of most Americans, the theology that we atheists are often criticized for criticizing. Most Americans do not believe in a fuzzy, vague sort of deity and a nebulous afterlife that involves becoming one with the cosmos. No, their god is a human-like entity with vast superpowers, and he sits on a throne with Jesus right next to him, and he is personally interested in everything we do. Most wonderfully of all, he listens to our prayers and does his best to answer them in a way that will make us better persons. He is a creator god who conjured us into existence himself; he is a wrathful god who will smite our nation if we disobey his rules; he is a loving god who thinks little children are just peachy and special. Paul Froese and Christopher Bader polled Americans and discovered that 47 percent of the population believes in a judgmental god who will decide your fate in an afterlife; another 24 percent believes in this same god but chooses to think he's a nice guy who'll let everyone into heaven eventually.

We shouldn't ignore these folks, even if they didn't attend seminary (although many of them did), even if they don't have *Ph.D.* after their names (some of them do), and even if they can't gas on endlessly with pretentious twisty jibber-jabber about the Divine Presence or whatever. We atheists have the courtesy to acknowledge their existence, at least enough to argue against their beliefs; unfortunately, sophisticated theologians prefer to dismissively pretend that the religion of the heartland, the simple belief in a literal anthropomorphic being named God or Jesus or Muhammad, while useful for keeping money flowing into the churches and Bible colleges, doesn't actually exist.

The substitution of an abstract proxy for the more literal myths most people hold is necessary. If the "sophisticated" theologians acknowledged the true nature of the Christian citizenry's religious understanding, they'd be far more concerned about their heretical, naïve, uninformed beliefs than they are about the minority of atheists. But they prefer to complain about offenses against the sacred made by outspoken atheists rather than address the more credulous and casual sacrileges of a populace that at least has the grace of believing in some deity.

What does the general public believe about their god? I have an advantage over priests in that regard: random people readily write or talk to me and argue for the existence of gods—so I get to hear from a lot of believers who enthusiastically and ferociously make their very best case for God to me, in hopes of converting me, or at least putting me in my place. These admittedly untrained theologians do a remarkably poor job of making a logical argument, so I must confirm the sentiment that most believers do not believe for rational reasons. They try, though, and I'll attempt to summarize their arguments for the truth of Christian belief.

The "you're going to hell" argument. Fear seems to be a major motivator for the religious people who contact me, because this is by far their most common tactic. They tell me gleeful tales about the horrible torments that await me after I die—the demons will pull me down and start flaying/burning/disemboweling/masticating me. The more grisly and intense their description, they seem to think, the more likely I am to see the light and flee into the arms of Jesus.

Unfortunately for their plans, I don't believe in hell at all, so ratcheting up the horror of a nonexistent place really doesn't make me tremble in my boots. It mainly tells me something about the psychology of my correspondents, who all seem to obsess over the agonizing fates of people who disagree with them.

Oh, and another peculiar thing about these correspondents: after dwelling on excruciating tortures that would put the bloodiest slasher movie to shame, they're most likely to sign off their letters with a cheery "Have a nice day," or "In Jesus' Loving Name," or something similarly incongruous. It might be sweet if they were trying to compensate for the horror of the Christian message, but I suspect that they are simply not thinking.

The "the Bible says it" argument. Speaking of oblivious correspondents, probably the second most common argument for the existence of God is the one that says the Bible is the one true holy word of God, and therefore it must be true, because God wouldn't lie. Only a fool could deny the existence of a god when the holy book of his faith plainly and inarguably says that the god of his faith exists.

Circularity and *tautology* are not words these people comprehend.

The most common Bible verse I'm sent is Psalm 14: "The fool hath said in his heart, There is no God." It's a little too self-serving of the Bible to be genuinely persuasive.

The "invisible pink elephant" argument. Many versions of the claim that you can't prove a negative are going around, but this is my favorite. You cannot make an absolute dogmatic assertion that pink elephants don't exist, because one might be hiding somewhere in the universe, and you haven't exhaustively searched every hidey-hole in the universe. Similarly, you can't reject the existence of any god because, hey, he might be hiding somewhere you haven't looked yet.

It's sort of true, after all. Because Christians are vague about where the deity is or what he looks like or what he does, I am effectively excluded from being able to say without reservation that a god of some sort might exist somewhere. But we can narrow the subject a bit and prove a negative. I am definitely, positively, absolutely certain that no pink elephants—defined as a large mammal of the family Elephantidae with a shoulder height of ten to fifteen feet and a weight on the order of ten tons, and with skin either painted or naturally pink—are present in my living room. Similarly, I can assert that Jesus, the son of God, Hebrew carpenter, and perpetrator of amazing magic tricks, is not in my living room.

At which time, of course, the evangelist will assure me that he is there, but he's invisible. He's also impalpable and silent, except for a mysterious inaudible voice that is supposed to emanate from my heart. We may as well give in and admit that there's also an invisible pink elephant gamboling in my living room, but that would be considered crazy.

Anyway, I think of these as arguments for the invisible pink guy in the sky. I'm not exactly clear on why I shouldn't think people who promote that fiction are crazy, too.

The argument from C. S. Lewis. Christians dearly love that gentleman from Oxford. He is probably the theologian most frequently referenced by my correspondents and interlocutors, and he may be the only one most of them have read—often in the form of his fantasy novels. In particular, they are fond of his trilemma: that the only conceivable explanations for the claims of Jesus are that he's a liar, a lunatic, or the lord; that obviously he's not a liar, because he was of such high moral character; that obviously he's not a lunatic, because we can find a few high-minded quotes from the man; so therefore he must have been what he says he was, the son of a god, of the same flesh as a god, and not just any god, but the divine creator and ruler of the entire universe.

It's a rather shaky leap of logic. The premises don't exhaust all the possibilities, the commonplace terrestrial explanations are dismissed on an extraordinarily thin rationale, and the conclusion is so absurd, such a vortex of illogic, that anyone accepting it must be as crazy as a poached egg, to use Lewis's own expression—which doesn't say anything favorable about its author.

I prefer to cite that other Oxford don J.R.R. Tolkien. Gandalf could only have been a fraud, which we know was not true because his fireworks were spectacular; or crazy, which we also know was not the case because he competently led the Fellowship through the mines of Moria; or he was a real wizard of the Maiar, a powerful spirit who came to Middle Earth in the Third Age. Therefore we know that Gandalf was a real wizard. Done.

Have fun. You can apply this same logic to deify any figure from history or literature.

The "but Christians are nice" argument. Usually this one is a matter of personal testimony; I could put together a whole volume of tales I've been told along the lines of "I used to be an atheist, and I was [strung out on drugs] [cruel to my family] [divorcing my wife] [etc.], but then I found Jesus and became a new man of high character and deep happiness, therefore Jesus was real." The entire churchgoing people of America must once have been raving angry atheist hedonists in broken relationships—which suggests that at an earlier time in our civic life, the parties were much more fun and the libertines far more common. Unfortunately, I've never been able to identify this magical period in recent history, even though I've lived through a few generations now. Yet all the Christians today seem to be citing this mythical past of ubiquitous godlessness. I really regret that I missed it all.

Having a conversion experience is not evidence of the validity of one's new belief. I'd go so far as to argue the opposite—that new beliefs founded in periods of deep stress and emotional trauma are not likely to be the product of reason and evidence.

The "first cause" argument. This one is rooted very deeply, and its proponents tend to be the most obtuse and most certain. They argue that everything must have a cause (which is usually where I stop them, if I can: they should show their work!), often telling me that this is a Scientifically Proven Fact. We can trace the chain of causality back to the first event in the history of the universe, which must therefore have had a cause, and that cause must be God. QED.

What they overlook is that their premise about causation, even if I accept it, isn't evidence of a deity. An avalanche—a massive, impressive event—may have had its origin in a minor, random event: the fall of a pinecone from a tree, the melting of a small patch of ice in the sun, a tremor in the mountain. None of these causes is an intelligent action, and none requires the intervention of something with even the brains of an ant. Perhaps the universe arose in a quantum fluctuation, a transient and entirely random surge in the vacuum energy of the cosmos. Is that God? I don't think so. Or perhaps our universe is simply an eternal super-universal substrate without beginning or end, requiring no initial cause at all.

It's also an awfully long reach from a putatively grand and glorious magical intelligence that consciously created our entire universe 14 billion years ago to the parochial little deity of the Abrahamic tradition who has temper tantrums over foreskins and fussily prepared food. The Christian God seems to have declined in majesty significantly from the good old days, before our ball of rock congealed out of the nebular gases.

The "the creator is self-evident" argument. I often hear this from people who oppose the idea of evolution, and it often involves prolonged, repeated use of the enchanted word *complexity.* Look at that amazing bunny rabbit; I can't make that, it's got an awesome bunch of parts that

I don't understand, and therefore a super-intelligent being must have created it. Isn't it obvious?

On the contrary, I think of this argument as obviously self-refuting. The bunny rabbit was made by a pair of bunny rabbits, creatures not renowned for their intelligence, and they put no conscious thought or planning into the conception, development, and delivery of their progeny. Nonthinking chemistry and biology can easily produce patterns of increasing complexity, so harping on how complicated something is doesn't help convince me of a history of intelligently assisted causality. Over the course of a semester, my office also tends to become increasingly complicated, with piles of books stacked on chairs and cabinets, papers drifting from desk to floor, and a growing accumulation of screaming memos demanding my attention. Most of that complexity is the product of neglect, not planning (especially the memos). From that I have to conclude that abandonment and laxity, not God, lead to growing complexity, and that if there ever was a god, the existence of elaborate phenomena like bunny rabbits is evidence that he's dead.

I'm hoping that publishing this short list of bad (but common) arguments here will lead to a decrease in their frequency in my mailbox. I have no illusions that I can possibly finally persuade their proponents that they're wrong, but just maybe a dim glimmer of a notion will appear in their heads that those arguments won't work on me, and that my critics need to be a little more creative. I would also appreciate it if a few of those theologians who complain that we atheists only tackle bad, stupid arguments, instead of their own sophisticated and powerful ones, would give me their addresses. I'd like to forward to them a colossal volume of correspondence so that they can spend their time productively correcting the misconceptions of a few hundred million deluded Christians, who are all under the impression that they're defending the existence of an invisible pink guy in the sky who we must believe in because he has a fantastically imaginative torture chamber.

The Karen Armstrong Diet

I am partial to the scientific, natural diet myself. I like amino acids, those nitrogen-rich building blocks for protein, the strong stuff that builds muscle and connective tissue. Carbohydrates? They're fast, flexible sources of energy, sugars and starches that keep me dancing and my brain purring along. And fats are pure rocket fuel, 39 kilojoules per gram. Perhaps they're a bit risky to one's health in the long run, but there's no better source of concentrated power—and they're also the key component of cell membranes. Whether I'm dining on McDonald's fast food or the finest haute cuisine, I'm going for the substance of matter.

There is another diet that's had a good run of completely incomprehensible popularity: the fundamentalist table. Forget the difficult chemistry of proteins and carbohydrates and lipids—fundy chefs will plop down a platter of flaming brimstone and flung stones, and the diners will like it or else, despite the fact that eating it will make you crazy or even kill you. The menu isn't so much a list of tasty and nutritious items as it is a madly scribbled extortion note. Don't worry, though, there are rewards: if you make it through the whole meal, you get a lump of hard candy and the privilege of meeting the cook—whom you will praise enthusiastically and loudly, forever, to distract him from the possibility of serving a second course.

Unless you've been brought up on a regimen of regular sulfurous fulminations and they're all you know, fundy fare isn't very appetizing and actually tends to drive all but the most dedicated (or insensitive) away. Those who want their meals seasoned with soulful spiritual fluff have to go elsewhere—but fortunately, there are plenty of other cuisines that have neither meat and potatoes nor acrid imprecations. They can turn to the Karen Armstrong Diet.

Karen Armstrong is a former Catholic nun who left the narrow and often unkind strictures of that faith to make a career as a commentator and author on religion. She has rediscovered religion as a nebulous source of vague meaning, which she makes excuses for in many long books that seem to consist mainly of earnest pleas for reaching for the meaningful in the meaningless. In the Karen Armstrong Diet, the only spice is transcendence. Transcendence, transcendence, transcendence. Human beings reach for something greater than themselves, have ineffable feelings of concentrated incomprehension, and presto, that can be distilled into a justification for religion.

Unlike the fundy greasy spoon, Armstrong doesn't demand that you unthinkingly worship the cook. In fact, she insists that the cook has no personality, no desires, no attributes that we would even recognize as belonging to a being. He doesn't speak or act—he's beyond mere language and the barbarity of physical action. That's a convenient property for him to have when he botches an order, since no one is going to be able to complain.

This god is so amazingly transcendent that even existence is beneath him. Expecting that great chef in the sky to actually exist is a primitive concept, in Armstrong's view, and worse, it's unspiritual and unreligious. As for those other ingredients of various faiths—such as an afterlife, a soul, the occasional miraculous intervention by the divine, a history of specific activities by the deity, a personal concern by the supernatural being for the believer, sectarian concerns—how crude. How unsophisticated. God doesn't actually exist in any sense of the word; he is above and beyond and more than those mundane ideas that drive billions of people to belief, and that have exercised the circumlocutory rhetoric of theologians for millennia. And of course, the most important tool in the Armstrong cookbook is the blender: all the differences between religious beliefs must be whipped into an airy, textureless froth, with all flavors lost, to find the truth.

The approach isn't entirely repulsive. I entirely agree that the whole concept of god is meaningless and pointless, and especially that the personal deity who finds lost car keys and damns people to hell if they have sex with the wrong person is definitely a primitive myth. I could almost buy into the Armstrong Diet if I weren't put off by its pretentious preciousness. The line between literally nonexistent and transcendently immaterial is a fine one—so thin that it actually doesn't exist.

Oh, excuse me: it's so thin that it transcends any tawdry preconception that it exists. It is beyond nonexistence. The distinction between nonexistence and invisible and imperceptible is so far gone that it has become numinous and sacred; someone somewhere is going to be tempted to worship this gap for the awesome immensity of its infinitesimalness.

I think Karen Armstrong already is.

If Karen Armstrong were to open a restaurant, she would immerse herself in study under the most famous chefs in Europe and Asia. She would be a wonderful scholar of food; she'd be poetic in her expression of awe, and she would praise every culinary tradition she encountered. She'd use her erudition to purify the nature of cookery down to its fundamental essence, and she'd write many books explaining exactly what that essence is, although we might be troubled by the fact that nothing she ever said seemed to be very concrete. She'd have an explanation for that, too—who wants to eat concrete? We want light, we want magic, we want an airy spiritual quality to our divine food.

Armstrong the restaurateur would be a master of presentation and atmosphere. Her tables would be gorgeously appointed and perfectly lit; the waitstaff would be impeccably polite and helpful. The calligraphy on the menus would be the most refined art, and you'd want to take them home and frame them. The service would be just right; it's not that there would be no waiting, but every unoccupied moment would be filled with contemplation and anticipation. The people who came to her restaurant would all be representatives of a social ideal—patient, kind, pleasant, and excellent conversationalists. Boors and the impatient wouldn't be excluded but would find themselves filled with soulful appreciation as soon as they stepped through the door.

Of course, there would be no kitchen. The Armstrong Diet transcends primitive notions of "food."

Eating would be replaced by therapeutic meditation. The best meals are spiritual exercises, like poetry or dance, that exalt us in their performance and provide us with a form of nutrition that is beyond mere chemistry—we would savor wonder and awe rather than the gross physical phenomena of flavor and full bellies. Why bother eating if we can instead think about what Maimonides and Aquinas wrote about food?

As Ms. Armstrong knows, she will have no shortage of clientele, no dearth of fawning rubes who will seek enlightenment in the profundity of nothing at all. They will at least be able to claim that they aren't just ignorant, they have moved beyond ignorance to a state of transcendent delusion. And they'll lose weight as well!

She won't be receiving my custom, I'm afraid. I'll stick to earthy sustenance that tastes good and fuels the body, even if the superb salmon and radicchio salad on my plate makes my brain look fat.

God's Little Crisis of Confidence

Once upon a time, God and his prophets were rather more assertive than they are now. God was all powerful, and proud of it—he wasn't at all shy about smiting, or sending howling hordes of Hebrews out to slaughter, whole tribes of heathens. Sodom and Gomorrah offended him, so—zap!—he annihilated those cities in blazing columns of all-consuming fire. He created everything on earth in just six days, because he wanted to, and when he got peevishly annoyed at people's wickedness, he killed every living thing (except for a few who could fit on a very big boat) in a flood. He was God, you know: omnipotent and able to do anything on a whim.

His prophets were just as fierce. When God destroyed the world with a flood, how did he do it? He just made it rain! Rain! It's that simple! Don't trifle with excuses, don't fuss over calculating how much water vapor could be present in the air, or estimate the volume of water required to immerse Mount Everest, and don't ask where all the water went afterward. God snapped his fingers and the deluge was falling out of the sky; then he snapped them again and the water was gone. He created the whole cosmos in an instant, so a planet's worth of water would be trivial. As for ye of little faith who calculate the mean volume of the standard antediluvian animal and make estimates about how many species could be wedged into a gopherwood ark, it's a waste of time; if a multitude can be fed with five loaves and two fishes, isn't it clear that the deity is not bound by space, time, mass, or closet space? God! Omnipotent! Two words that go together and explain everything!

Self-respecting patriarchs and prophets don't need rationalizations—they realize that restraining their god to material mechanisms is limiting. But nowadays we have a shortage of true believers in an omnipotent god. I'm not speaking of godless atheists: we just dismiss the whole story as a funny fable entirely lacking in justifications. No, I'm referring to the creationists, the ones who take the Bible most literally.

The creationist literature since Henry Morris and John Whitcomb, in the early 1960s, has been a pathetic plea for materialist credibility. They don't quite trust the all-powerful agent at the heart of their legends. It's not enough to say he could simply do anything by a trivial exertion of his will. No, he must follow rules.

Consider the case of the creationist Walt Brown and his hydroplate theory. He explains the flood and the sudden appearance of a deluge of water by postulating that (1) the earth's pre-flood atmosphere was supercharged with oxygen, and (2) the earth was surrounded by a firmament, a barrier of solid metallic hydrogen. To cause the flood, God allowed the oxygen and hydrogen to react, diminishing the oxygen content of our air and vaporizing the hydrogen canopy, but producing a huge quantity of H_2O to drown the planet.

It's a bit overwrought, don't you think? Here, allow me to channel my inner Abrahamic patriarch: *You silly git, you're thinking too much! God can do anything! He just made it rain! Poof, rain! Miracle, get it?*

But Brown's story does leave something miraculous for God to do. His awesome detonation of hydrogen and oxygen would be a highly exothermic reaction. It wouldn't be rain coming down, but superheated steam that would sterilize the whole planet like a giant autoclave. So we'd need some magic to keep Noah's big boat from being cooked. It's not that God couldn't do that (*He can do anything!*), but that it's decidedly unbiblical. It rained for forty days and nights, remember.

Walt Brown isn't alone—the whole so-called scientific creationism movement is in the same awkward situation of trying to get their Old Testament superhero to adhere to the rules of physics, chemistry, biology, and ordinary common sense—which he is still free to violate whenever his apologists have backed him too far into a corner.

In Kentucky, for instance, Ken Ham has spent tens of millions of dollars building the Creation Museum, which tries to ape genuine natural history museums. It's not a church, not a cathedral, but a museum filled with poorly organized exhibits and fossils and animatronic dinosaurs frolicking with animatronic Adam and Eve, all intended to prop up an inadequate faith with scientific support. But if visitors already believe, if their faith cannot be contradicted by mere physical evidence, why do they need reassurance that the rocks don't defy their chronology?

This Creation Museum presents a whole series of facile but implausible efforts to resolve the difficulties in postulating a creation event that occurred only six thousand years ago and a near-total extinction event that took place in 2348 BC. For example, the amount of radioactive decay that has occurred indicates the earth is far older than six thousand years. How do they account for that discrepancy? Radioactive materials must have decayed incredibly rapidly in biblical times. How did koala bears and New World monkeys disembark from a boat in the Middle East and end up in Australia and South America, respectively, in a short period of time, without getting mixed up? They hitched rides on rafts of floating logs. Why did *T. rex* have such large sharp teeth when animals before the Fall didn't die or eat each other? The better to eat fruit. Who did Cain marry, when his was only the second generation of humans to exist in the world? His sister, but it was all right because people hadn't yet accumulated deleterious mutations.

The accumulated excuses are mind-numbing. Why not just declare that God created the products of radioactive decay in the rocks? Poof! He can make lead just as easily as he can uranium. Why not say that God dispatched a team of angels to shuffle the various species from the grounded ark to the destinations he desired? Why not simply declare teeth and genetics irrelevant? God, once upon a time, would simply have conjured a nutritious diet directly into the guts of his creations or, when it came time for his closely related creations to breed, whip up a custom genome on his handy-dandy DNA synthesizer and swap the nonincestuous products into the sperm and egg. *He can do anything,* remember. We shouldn't need to add modern technical details to Bronze Age fairy tales—unless perhaps the devout listeners to these stories need just a bit more verisimilitude before they can swallow them. Without a leavening of science, after all, religious fables are simply unbelievable.

It gets worse. The latest breed of creationists, the proponents of intelligent design, have fled so far from their convictions that they hide the bearded old sky god from the public eye and make him a silent conspirator who sneakily and undetectably (but indispensably!) nudges his creations over hurdles that are too high, in their estimation, for natural processes. They avoid mentioning the old man in public; it's safer to speculate that maybe aliens were involved, or nonsectarian magic, or that quaint old myths like astrology are reasonable alternatives to science. It's almost as if they're embarrassed by the nature of their designer.

Or look to the theistic evolutionists for some serious trivializing of their god. Ken Miller, in his book *Finding Darwin's God,* reduces the omnipotent, omniscient lord of the universe to a manipulator of quantum indeterminacy—he exercises his awesome will in gently determining precisely when an atomic particle will decay, or nudging the vector of an emitted photon—all of which is completely undetectable. Darwin's god is apparently a penny ante card sharp who likes to shave the odds and, at the same time, demand that people who can't detect what he's up to worship the invisibility of his skills.

It's a good trick, but what we're seeing is the progressive enfeeblement of God in the hands of his believers—who still desperately insist that his existence is a certainty. Don't blame atheists for the long, slow strangling of this weak and incompetent deity—we're offering the swift relief of a mercy killing! If God were real, reduced from cosmos conjuring and planet smashing in his youth to barely manageable electron diddling in his old age, he'd probably appreciate the kindness of the atheists' pulling the plug on him.

Laughter as a Strategy for Diminishing Religion

There's a time to take religion seriously and a time to laugh at it. Sometimes we have to do both at once.

When you read the works of apologists for religion, you find they share a common tone: they're all very serious. Religion is important. Faith is about our relationship with the greatest being in the universe. We need God to find purpose in our lives. Religion is about community and bringing people together. Rituals are tools to cope with the great events in life, from birth to death.

It's all true in the sense that believers really do believe that their god matters, and sometimes the greatest kindness that we godless give to them is that we take them seriously and argue seriously about the meaning of their beliefs. We engage them on their chosen turf and tell them they're wrong, and present the logic of atheism, and point out the failings of religious belief, and note why uncritical, irrational thinking hurts our culture—and we have to do that, because this is a serious debate about how the universe actually works and how society should accommodate myth and superstition and common delusions.

But the other side of the story is that, boy, religions sure do promote some goofy stuff that is so ridiculous it doesn't even make convincing comedy. The absurdity is so ripe that we often pass over it—surely they must know how silly that belief is, right? They don't really hold that bizarre position—this is supposed to be a serious discussion. But they do. And we should not ignore the power of laughter.

In 2009, when people were deeply concerned about the swine flu pandemic, a flock of fifty rabbis took to the air and flew around Israel. As their plane patrolled the borders, these very serious men in their distinguished beards and yarmulkes and somber clothing shouted invocations to Adonai and read their holy books and blew on magic horns to stop a virus. It was madness. It was ridiculous. It was utterly ineffectual.

There are multiple ways to respond to such absurdity. Do the science. Measure infection rates before and after the shofar blowing. Do many trials. Analyze the statistics. Publish the results in an epidemiology journal. Good luck with that—it's important to actually test the claims, but also very hard to do, and the research is also unlikely to meet any credible standards for publication. (The use of shofars to control viruses is not an active controversy in the biomedical community.) The hypothesis is also going to take time, money, and effort to test properly. If done well, the study might be useful in persuading some rational rabbis in the future that chanting and howling in the sky is not a smart strategy to ward off disease. But that finding is also going to be ignored by the majority of rabbis who do not read the scientific literature and simply don't care about the data, and most will rationalize their way out of the logical rebuke by saying: "We're not dying of swine flu now, so see? It worked!"

There's another way to respond to the absurdity. Show your audience a video of the crazy rabbis trying to go medieval on a virus from a modern airplane, and let them laugh. Now, instead of having to teach them the statistical methods of epidemiology and how the shofar method failed to conform to them, all you have to do is ask whether they really want to be associated with that carnival of lunacy. It's easy! And cheap! And it builds on a basic principle of human nature: people don't want to be ridiculed. Humor is a far more potent crowbar than statistics for wrenching people from their preconceptions.

Have no illusions: neither approach is universally effective, and neither will have a tremendous success rate. But then, as far as I know, no one has ever done a controlled study of shofar effectiveness, so that isn't even an option. And all it takes to make a net gain is for a few people to be embarrassed by their flying rabbis and then go ask their doctors about the best way to avoid getting the flu.

Here's another example of a religious absurdity. In 2010 Hojatoleslam Kazem Seddiqi, an Iranian cleric, announced that women dressing immodestly cause earthquakes. He even used the fear of divine wrath to threaten Iranian people with death "under the rubble" if they didn't get on board with the theocratic policy of oppressing women. It's laughable on its face. How would such a phenomenon work? Where is the evidence? Isn't it obvious that local mores about dress differ from place to place and aren't correlated with earthquake frequency?

Once again, we have different ways of responding to this

claim. One is to calmly and rationally point out the qualitative evidence and even go through the historical record and do a detailed analysis. We could note that repressive Iran, where the women are covered up heavily, seems to be rather more prone to seismic wrath than, say, the beaches of the French Riviera, where women may wear nothing at all. But let's be honest: that bloodless approach, effective as it may be, doesn't excite the imagination. We could drone on and on for ages, and no one would notice. And as scientifically strong as a message may be, it's useless if it isn't heard.

Jennifer McCreight did something different: she called for a boobquake, suggesting that women dress immodestly (by Iranian cleric standards) on April 26, and thereby invite God to smite the planet with earthquakes—a call that could only be made in confidence that Seddiqi's claim was baseless. And the boobquake was a phenomenon. It was picked up by news media around the world, got millions to pay attention, and effectively highlighted the silliness of a religious claim. The event was media-savvy and human-psychology-savvy—it used humor, sex, and fun to make a serious point interesting, and it led people to look at the science of earthquakes. Did it make hordes of Muslims convert to atheism? Of course not. But it did make an Islamic authority look a bit more ridiculous in the eyes of the world.

Religion has at least two weaknesses. One is that it is empirically false. All its specific claims are either pointless and unverifiable or else have been falsified. The other weakness, one we neglect at a cost of diminished effectiveness, is that religion is hilarious. It's a prime target for exposure of folly; we need more people to exploit the soft, ticklish underbelly of faith.

Unfortunately, right now, the atheist community is needlessly split between two poles. On one side are the softies, who complain that believers don't deserve ridicule, that hard truths and blunt speech and laughing at fervently held beliefs simply hardens people's hearts and drives them away. We have to be sensitive and avoid confrontation, they assert; logic and gentle persuasion will win the day. On the other side are the hard-edged ones (the current favored term for them is *dicks*), who point out that you can't reason someone out of a position he didn't reason himself into, and meanwhile their fond religious beliefs are being used to hurt people, and so they must be strongly criticized and mocked. And really, religion is a clown circus, and asking us not to point and laugh is unnatural and dishonest.

Both sides are wrong, and both sides are right, and there sure aren't many people standing at either extreme. You can reason some people out of indoctrination, and slow and patient instruction can win people over to atheism. I know some of them; they write to me and tell me that something I said actually led them to think through their position. But shock also works. Ultimately, people hold their religious beliefs for emotional reasons; deep down, fear and comfort, disgust and empathy are the tools religion uses to manipulate natural human desires. We would be idiots to shun emotional appeals, and it would also play into the ridiculous Spock stereotype of atheists as cold, dead, soulless people who substitute math for passion.

Sometimes you can reason people out of deeply held beliefs. But it helps if first you stir discontent, if you wake them up to the fact that their beliefs make them look ridiculous—and that, yes, a whole group of people are laughing at them.

Making believers and belief the butt of the joke is another form of sacrilege—and oh, they do hate that. It's an entirely human response—so use it.

We're Happier out of a Straitjacket than in One

A while back I saw something wonderful at a science-fiction convention. People often put on odd and extravagant costumes at these events, and one rather obese young man made a minimalist choice: he came as one of the Spartans from the movie *300,* which meant he was standing in the crowd wearing a red Speedo and a bright red cape—and nothing else.

Now imagine this same young fellow at an event at your high school. The reaction would have been brutal. I know; when I was in high school, I was a little Poindexter, ostracized, laughed at, and treated like a space alien. But my ostracism was relatively mild: those who were even more different—the fat kid or the gay kid or the homely kid or anything else that excluded you from the jock clique or the Heathers or whatever ideal the majority of the student body worshipped—were mercilessly tormented with unremitting cruelty. Often in our culture, socialization is achieved by maltreatment and unkindness based on superficial differences.

But this fellow at the convention, he was smiling and laughing and having a good time. Other people were smiling back and complimenting him on his costume—he was being rewarded, not for having a jock's body, but for being open, a bit ironic, and unashamed of who he was. And that was a beautiful thing. He wasn't alone, either; these events are places where people who are misfits in the conventional world are free to blossom and be themselves, and being different is valued.

This doesn't mean that con culture is value-free and open to everything, or that it is entirely anarchic: rather, it means that the group's ideals are focused on properties of the mind, in particular creativity, imagination, and boldness. Here conformity is the sin. Here rules that infringe on freedom of expression, that condemn someone for who he or she is, are the transgressions.

I was reminded of this mind-set by events at the 2010 San Diego Comic-Con, where the hateful fundamentalists of Fred Phelps's Westboro Baptist Church showed up to do their usual stunts and to damn everyone who doesn't follow the strict rules of their narrow interpretation of the Bible. They are the antithesis of what I appreciate about cons, so it was like a matter-antimatter collision on the public streets. But the wonderful thing was that the con attendees responded wonderfully, laughing and joking and being flamboyant. They staged a counterdemonstration, showing up in costume and waving signs saying "Odin is God—Read Mighty Thor #5" and "God Hates Sentries," among many cryptic references to obscure comic book tropes. This was also a beautiful thing.

The other commendable fact of this event, as I know because I know these kinds of people well, is that if the

Westboro Baptist picketers had walked over to the Comic-Con counterprotest and said, "Let's talk," the con crowd would have cheered and welcomed them and enthusiastically joined in conversation—because the WBC people are weird, and the con people love to embrace strangeness.

Unfortunately, the members of the Westboro Baptist Church couldn't do that. One of the terrifying properties of fundamentalist religions is that they demand conformity: they lay down strict rules to regulate how people are allowed to think and behave and often even dress and speak. Questioning the dogma is forbidden. Situations where they have to think for themselves are dangerous. The community must be policed so that odd notions do not pollute the minds of their children or themselves.

Perhaps the cruelest aspect of conservative religions is their insistence that all people must follow one straight and narrow path, regardless of the fact that people are diverse. By fostering fear of deviation, they impose endless misery on so many people. It's the Chinese foot-binding of the human mind.

Now I'm not trying to imply that the wild and crazy crowd is free of all dogma: the caricature of Comic Book Guy on *The Simpsons* is an excellent example. Sometimes even the oddballs can get caught up in the details of their fantasy canon and forget to have fun while carping over fidelity to an ideal. In this context, I can't condemn all religions, because in the wild and crazy crowd, you'll find many members of liberal religions, many people with strange ideas about spirituality, and more than a few techno-utopians. If you're going to encourage weirdness, weird ideas will flourish.

And that's all right. As long as no one is forcing others to accept Jesus as Lord, or that *Star Wars* is better than *Star Trek,* then we're doing fine. The human mind should be a playground, not a military camp.

I do confess to some bias. I've chosen academics as my career, which has some significant strictures on behavior. (I could not show up to teach a class in a Speedo and a red cape—probably.) But it at least offers a great deal of intellectual freedom. I also grew up in the 1960s and 1970s, when—well, you had to be there . . .

This generation, or some generation soon, I'd like to see the hippies win. Peace and freedom, man.

Besides, it makes the fundies and Republicans freak out.

Marketing Godless Science

I am afflicted with an unfortunate allergy to a common strategy for communicating to the public called *framing*. I believe it may take its name from the practice of building wooden frames on which to crucify those who don't follow its precepts, or perhaps it's taken directly from the colloquial term for falsely incriminating someone. One odd thing about framing is that its most avid proponents are absolutely miserable at it, and as my first piece of evidence, I cite the fact that they named their practice with a word that is more commonly found in pulp-crime novels, and it's something the bad guys do. At any rate, I break out in angry hives whenever I hear the word.

I shall try to be fair anyway. It actually is a legitimate idea, sanctioned by social science, that one should research and understand one's audience and tailor one's message to appeal to their best interests, drawing on shared values to make a connection and generate sympathy for one's cause. It's great if you're selling soap or trying to convince someone to vote for you.

Unfortunately, I have no soap to sell.

But I don't want to give the impression that I deny the utility of social science research. Far from it; psychology and sociology are genuinely scientific fields of study that bring discipline to the analysis of the subjective. I've read research about how people respond to various cues, and what kinds of communication tactics are actually persuasive to others. I think the investigators are honest, are reporting their data accurately, and do reveal interesting facts about how the human mind works. I just don't believe it applies to the conflict with religion, because trying to encourage people to think for themselves is a different process from trying to get them to become a sympathetic follower; because manipulating people by pushing the right psychological buttons may work in the short term but not necessarily in the long run; and because, well, I feel dirty if I hide my real views in order to better tweak people into liking me and my message.

Attempts to find ways to manipulate the subjective have a weakness, too—the manipulated may become aware of the manipulations.

For example, a study by Bruce Rind and Prashant Bordia in 1996 found that waitresses could increase the size of their tips by a whopping 28 to 33 percent by drawing a smiley face on the check they'd give customers. Phenomenal! If you're a waitress trying to make a living off low wages and tips, you'd be crazy not to take a few seconds to scribble something like that on the check. Immediately after reading about that study, my wife and I started noticing, at restaurant after restaurant, that our waitresses were drawing smiley faces all the time. (And now you'll notice as well.) I doubt that most of them had read the academic article, but versions of this story had made it to the popular press, and it was lovely to see people applying empirical research to their work.

Except, alas, my own perception shifted. Now when I look at the back of that check and see the standard iconographic smiley face, I'm wondering if I'm being psychologically manipulated: Do two dots and a curved line really make up for poor service? I'm getting inoculated against the positive response, and the sight of a smiley face makes me think instead about what the waitress actually did. And then what if there is no smiley face drawn? Does that mean my waitress is unhappy in her situation, and maybe I should tip a little more in sympathy? Or perhaps I should express my regret at a bad system that justifies low wages for hard work. The reward response for a simple cue crumbles very easily.

Costly signals are always more reliable indicators of honest communication than these cheap and easy markers. The ubiquity of push-button pop psychology can lead only to an increase in suspicion and critical thinking, which is, perversely, a good thing, I suppose. We don't like insincerity. We're wise to question cheap and easy answers.

Imagine encountering a man smiling confidently.

Q Why are you smiling?
A Because market research says I should.

It backfires right away. We don't trust that man one bit.

But wait, you say—his answer undermines the signal he gives! Here's a more realistic version of that encounter, then.

Q Why are you smiling?
A Because I'm happy and confident and someone you should listen to!
Q Really?
A No, but that's the answer market research says I should give.

There are a great many formulas for short-term success, and many of them are backed up by sound psychological research; many of them are also backed up by the pop-psych prosperity gospel on display on late-night infomercials, which use strategies that are successful in winning profits for their makers—for a while, before they have to move on to the next moneymaking fad. I would not argue with demonstrable success or replicable data.

But I do question the objectives of such exercises. I'm not trying to win people over with glib patter, a happy smile, or sex appeal, and neither are any of the prominent voices of the New Atheist movement. We're out to unsettle people, make them think, make them question, and sometimes make them angry—so that they will, we hope, think more deeply themselves and come to their own conclusions.

I have no testimonials from people saying I brought them around to my way of thinking with a sharp, cogent argument. That's simply not the way the human mind works when confronted with challenging ideas. I have many letters from people saying that I pissed them off, that I made them look up the evidence to show that I was wrong, and that eventually they changed their minds and came around to something close to what I'd argued—but never exactly right on to my views. That's the curse of skeptical and scientific thinking: it doesn't produce sheep, only fellow wolves who will be better able to argue a little more fiercely with you.

And that's the way I want it.

“Science Is What We Do to Keep from Lying to Ourselves”

Science and religion are two different ways of looking at the universe and changing the world, and I believe that you must set one aside to follow the other. One works; the other doesn’t. One struggles to expose the truth; the other conspires to bury it under a burden of myth. One is a method of analysis and experiment; the other is pretense and lies.

There is vehement disagreement on this issue, though. Most of the opposition comes from people who are both religious and scientific, who are eager to make the most common—and also the most inane—argument for compatibility. This argument consists of squeaking, “It’s not true! I’m a scientist and I go to church every week!” or of listing all the people one knows who are both scientists and Christians, such as Ken Miller and Francis Collins, who are definitely scientists and most definitely vocal Christians.

I would not disagree with their details: there are many good scientists who believe in a god or gods, and I’m certainly not going to imply that someone who is a churchgoer and believer can't be a scientist—he or she certainly can.

Here's the problem with that particular pro-compatibility argument. Let me tell you about Dennis Rader. He's an evil man, more notoriously known as the BTK Strangler, BTK standing for Bind-Torture-Kill. He murdered at least ten people in the American Midwest between 1974 and 1991, and he wrote ghastly gloating letters to the police, taunting them and bragging about the vile acts he committed. He's an unforgivable psychopath, and fortunately he's now locked away deep in a federal penitentiary, where we all hope he will rot.

He's also a Christian. He was a devout member of Christ Lutheran Church in Wichita, Kansas, for thirty years and was elected president of the congregation council. He was also a Cub Scout and Boy Scout leader.

Which leads one to ask: Is serial killing compatible with Christianity?

Now you might be mentally reviewing the history of Christianity, picturing crusades and inquisitions and pogroms, and thinking, "Uh, YEAH!" But let's be charitable. At least Christianity has gradually moderated its intolerance over the centuries (one could argue that Islam hasn't reached that level of maturity, or senescence, yet), and nowadays it's not quite as rabid as it used to be. (Which reminds me of the best definition of a New Atheist I've heard: it's an old atheist whom the Catholic church can't legally set on fire anymore.)

I think, honestly, that we'd agree: serial killing is incompatible with modern Christian thought, and it's also incompatible with the high-minded ideals of the religion.

Christians would even more enthusiastically agree that Dennis Rader was not acting as an agent of Christian values. The fact that a few deranged individuals are Christian, or even that the majority of criminals in the United States are also Christian, doesn't mean that the philosophy and values of Christianity endorse criminality in any way.

But we rarely observe a serial killer who squeaks that he is a good Christian and who also tortures a prostitute every week. Likewise, serial killers who were also Cub Scouts rarely vouch for compatibility.

Let me give you another example from a different angle. Kary Mullis is a very well-known scientist: he discovered PCR, the polymerase chain reaction, a technique for rapidly amplifying known DNA sequences in a sample. It was a brilliant insight—PCR machines are ubiquitous in molecular biology labs all around the world, and he won a Nobel for it.

Mullis is also a certifiable kook. He wrote an autobiography in which he lists all the recreational drugs he takes, talks about the women he's had sex with, explains that HIV doesn't cause AIDS and that ESP is possible, and discloses that one afternoon he ran some data through a computer and proved that astrology actually works. It's wonderfully entertaining, in a trippy train-wreck sort of way. Now he travels around telling people that global warming isn't happening.

Is Kary Mullis a scientist? Yes, definitely. He has a Nobel Prize. PCR genuinely was a great scientific discovery. Does that imply that every thought that scampers through his well-doped brain is scientific? No, most definitely not.

Is astrology compatible with science because a scientist believes he has scientifically demonstrated it? No.

Our Brains Are Full of Contradictions

People are complicated animals. Our brains hold all kinds of false and contradictory ideas, and different people have different ideas—imagine that!—and those ideas have to be evaluated independently of the individuals holding them, as do whatever other ideas are tumbling about in the grab bag of their craniums.

When we claim that, because so-and-so is a scientist, every thought in his head must be automatically legitimate in the scientific sphere, we're making a glorified argument from authority—we're making a logical fallacy. There's plenty of room for each one of us to be wrong. Also, this defense of scientific compatibility, I've noticed, is always brought up to support religion and nothing else. If I tried to argue that my fondness for cheesy science-fiction movies like *Mega Shark vs Giant Octopus* is a validation of their accuracy because I AM A SCIENTIST, I don't think I'd get very far.

Religion is the Mega Shark of culture. It's improbable nonsense that tickles our brains in amusing ways. It's an unreliable and outright falsified collection of silly stories. And the CGI really sucks, too, the acting is wooden, and the story is unlikely.

The part of religion that is open to criticism, that makes it incompatible with science, is that it is demonstrably wrong, and further, that it is remarkably refractory to correction. Religious apologists can be brilliant people who build the most amazingly intricate rationalizations for whatever weird practice that they favor—whether it is ritual cannibalism or praying to spirits or treating women as chattel—but they always build on a shaky foundation that is either falsifiable or unsupported by evidence. The justifications they construct can lead to institutionalized charities, which are a good thing, or to tribalism and holy war, which are generally bad things—and either is about equally supported when your source of authority is an invisible man in the sky who whispers inaudible orders directly into the brains of his priests.

Where religion clashes with science is that religion makes smart people do stupid things. And scientists shouldn't like stupid.

Ken Miller, Poster Child for Compatibility

Ken Miller is an American scientist, a cell biologist, and a really smart guy. He's also an excellent speaker, and he's been a very hardworking asset combating creationism in the United States. I'm going to criticize the fellow strongly here, but let me emphasize: I want Ken Miller to keep doing most of what he's been doing, because he's a net benefit to the cause of science education.

However, Miller is also a Catholic. In itself, that's not a denial of his abilities as a scientist—remember, people are complicated and can hold all kinds of contradictions in their heads. But it does mean he believes in some deeply silly things, and unfortunately, he's the poster child for the claim that science and religion are compatible. Which means I have to deal all the time with people who want to argue that Catholic dogma is scientific.

Here's a specific example. Ken Miller is a biologist, and he's familiar with the evidence for evolution, and fortunately Catholicism has no official complaint with evolution. Much of Miller's popular writings are all about why evolution is a fact, and he's very good at knocking down the follies of creationists.

See? Smart guy.

But problems emerge whenever he tries to reconcile evolution, which is an impersonal force with some rather heartless and cruel implications, with his Catholic belief in a loving god who intentionally created us. There is no scientific evidence for such a being. There is no evidence that we are the product of anything but chance funneled through the filter of pitiless selection. Miller knows this—so he had to invent a loophole to wedge his god somewhere into the process, in a way that excuses the absence of evidence.

He makes what some call "explicitly scientific arguments" that God is found in quantum indeterminacy: because the behavior of subatomic particles isn't fully predictable, God can "intervene in subtle, undetectable ways." That is, the dice that God is playing with are loaded. And whenever we try to scrutinize the process, he's really, really good at putting in a fair set of dice. God cheats at physics.

Terry Pratchett, the fantasy author, says, "Let's call it Quantum!" But that's not an explanation.

Can I disprove Miller's assertion? No. He has carefully crafted his explanation to be completely untestable. But as anyone with even a passing familiarity of science can tell you, untestable explanations with no supporting evidence are not, despite Miller's insistence, scientific at all. That is an entirely religious argument, by which I mean it is really a fact-free argument. For one thing, how does Ken Miller know that his god nudged an occasional cosmic ray into bumping nucleotides in our ancestors' DNA? No one has observed such a thing, and Miller is openly admitting that it is unobservable. Yet the history of mutational change is fully compatible with an alternative explanation that these changes were driven by purely chance events—and that alternative has experimental confirmation, unlike Miller's completely speculative scenario.

It's as if Miller hasn't read what should be everyone's sourcebook for the principles of critical thinking, Carl Sagan's *The Demon-Haunted World: Science as a Candle in the Dark*. I strongly recommend that he read the chapter titled "The Dragon in My Garage," which is precisely the story that Miller is telling us, without any quantum. If you haven't read it already, run out and buy a copy. To summarize, he tells a little fable about keeping a fire-breathing dragon in his garage. Except that it's invisible. And it floats so it doesn't leave tracks. And its fire is heatless. And it's not corporeal, so you can't touch it. It's a lot like God, constantly removed from the realm of reality.

Sagan's conclusion is the smart one, and it applies to how we should treat God and Miller's little quanta.

> The only sensible approach is tentatively to reject the dragon hypothesis, to be open to future physical data, and to wonder what the cause might be that so many apparently sane and sober people share the same strange delusion.

I am not through with Miller, though. In a 2010 magazine interview with David Scharfenberg, he compounded the poverty of his explanation. Miller justified the subtle and undetectable manipulations of his god by suggesting that while God could regrow limbs, he wouldn't do so because it was too easy, turning prayer into a way to control God's behavior.

> Suppose that it was common knowledge that if you were a righteous person and of great faith and prayed deeply, all of a sudden, your limb would grow back. That would reduce God to a kind of supranatural force . . . and by pushing the button labeled "prayer," you could accomplish anything you wanted. What would that do to moral independence?

It leaves one a little breathless. As I said, Ken Miller is a scientist and a smart fellow, but religion leads him to make ridiculous leaps of illogic. His argument for why God doesn't heal amputees reveals that he hasn't read another important work in the scientific literature, Douglas Adams's *The Hitchhiker's Guide to the Galaxy*. I'm sure most of you recognize Miller's argument as simply a variant of the Babel fish story. The Babel fish is an incredibly improbable organism, a small fish that you can stick in your ear and that uses telepathy to act as a universal translator.

> Now it is such a bizarrely improbable coincidence that anything so mind-bogglingly useful [as the Babel fish] could have evolved purely by chance that some thinkers have chosen to see it as a final and clinching proof of the NON-existence of God.
>
> The argument goes like this:
>
> "I refuse to prove that I exist," says God, "for proof denies faith, and without faith I am nothing."
>
> "But," says Man, "the Babel fish is a dead giveaway, isn't it? It could not have evolved by chance. It proves you exist, and so therefore, by your own arguments, you don't. QED."

> "Oh dear," says God, "I hadn't thought of that," and promptly disappears in a puff of logic.

According to Ken Miller, then, there are no human amputees receiving miraculous recoveries at Lourdes for the same reason there isn't actually a Babel fish—because if God actually did something useful, it would prove he doesn't exist. This is a god so powerful, so awesome, that he refuses to be bound by the arbitrary requirement of making a goddamned difference in the world.

I have a couple of other problems with Miller's argument.

Many people have done research on amphibian regeneration—it's the stuff of undergraduate developmental biology classes. You can take a newt and snip off a whole leg with a pair of scissors; then wait a few weeks to months, and it grows back. I've done this several times, and I never noticed that I was disproving God's existence. Perhaps that's why I'm an atheist now, and all I have to do is get Miller into a newt lab to turn him into an atheist, too.

Of course, there are other alternatives: it may be that God just loves newts more than he loves us. Or perhaps newt wish fulfillment has destroyed their moral independence. Clearly, we need more research.

The other peculiar thing about Miller's argument that God works undetectably—that he can't be too obvious about miracle making because he might undermine faith—is that he's a Catholic. He believes that God turned into a man and wandered about in the Middle East turning water into wine, walking on water, curing leprosy, and raising the dead. Doesn't that make God a "supranatural force" right there? Does belief in Catholicism destroy one's moral independence?

Oh, if only God could vanish in a puff of logic, we'd be all done here!

But he doesn't vanish. Why not? Because the human mind is a product of evolution, an assortment of shortcuts and utilitarian routines that are just good enough to allow us to survive and thrive. It evolved to help us to eat and avoid being eaten, to keep us mostly functioning cooperatively with other members of our tribe. It is not a remorseless engine of flawless logic. We will, for instance, accept all kinds of incredible nonsense if it will get us laid, and we all know this.

We will even join a church if an attractive object of our lust asks us to.

Religion Fails as a Source of Knowledge

Religions are the lies we tell ourselves to justify our biases. Rationalizations, which are all that religion can generate, do not represent new knowledge; at best, religion generates comforting noises to mask the fact that someone is starving to death, or is dying of cholera, or has had a crusader stick a sword in his or her chest. That is all it has accomplished in millennia of human history—it has not improved our ability to grow food or to treat disease; nor has it given us new insight into our origins and existence. You would think an omniscient and omnipotent god, when dictating his holy book to his prophets, could have thrown in a few accurate suggestions that would have shown that he actually knows a bit more than the goat herders he was talking to.

For instance, Jehovah gives us an account of the history of his creation that is only about a page and a half long, is a bit poetic and repetitive, and gets everything completely wrong. It's a real waste. Why couldn't this all-knowing being have dispensed with the metaphor and poetry and written something like this as the first paragraph of Genesis?

> As many more individuals of each species are born than can possibly survive; and as, consequently, there is a frequently recurring struggle for existence, it follows that any being, if it vary however slightly in any manner profitable to itself, under the complex and sometimes varying conditions of life, will have a better chance of surviving, and thus be naturally selected. From the strong principle of inheritance, any selected variety will tend to propagate its new and modified form.

That's a brilliant insight, beautifully expressed, and it's short and doesn't use any jargon. Translated into the right language, I don't see why a Mesopotamian goat herder couldn't have appreciated it or, at the very least, come to understand it with contemplation and experiment.

Instead, the human race had to wait until 1859, when Charles Darwin wrote those words, to actually learn something deep and profound about the process that creates biological diversity.

Or if you think evolution would be a little too deep for Iron Age goat herders, God could have at least told them "Wash your hands!" as a basic and useful bit of hygienic advice that would have saved many lives. But no, we had to wait until the nineteenth century for doctors to figure that one out. Instead, we've got a Bible that gives detailed orders to snip off the ends of our penises, which should win a prize for the most useless piece of medical advice ever.

Science as a Lever to Move the World

I'm sure you've all heard that quote from Archimedes: "Give me a place to stand and a lever, and I will move the world." Archimedes knew that his merely human muscles were weak and inadequate to the task of moving a whole planet; we can't lift a single mountain, or a large boulder, or even our overweight neighbor. But he also knew that simple machines amplify the amount of work we can do. We use tools to increase our power beyond the bounds of what evolution has provided us.

Science is a tool, too. It is a process that helps us to overcome the logical shortcomings of our brains; our heads are full of models of the world that are useful in the short term, such as theory of mind, but are not necessarily accurate. They are good enough, but not necessarily true. The real world is often counterintuitive.

Richard Feynman had an excellent short description of what science actually is: "science is what we do to keep from lying to ourselves." It's a way of thinking in which we are constantly aware that we could be wrong in our preconceptions, and so we continually define new tests of our ideas. It is a way of breaking out of faulty models that evolution has granted us.

What has most improved human life in the last few centuries? Unquestionably it is new technologies, generated by this recently discovered process called science, a powerful way of knowing that really does create lots of new knowledge. We've tested the scientific tool and found it remarkably effective. We've also tested the religion tool and found that it has led to stasis and stagnation, ignorance and the worship of dogma.

Science is directly in opposition to religion as a way of knowing. Faith is not information, and belief is not sufficient to probe the universe—they are antithetical to critical examination and honest understanding. Adding the taint of religion to science doesn't complement our knowledge; it never has to date, anyway. And given that every example of religious apologetics combined with science that I've seen so far—from Ken Miller to Francis Collins to the fantasies of reconciliation of the Templeton Foundation—has merely cluttered up our understanding with nuggets of pointless myth that eventually have to be tediously extracted, I don't see that it ever will.

Science and religion are incompatible in all the ways that count: Science works. Religion doesn't.

The Active Hand

The more skilled the slave, the more likely he or she is to break free. Where faith enlists knowledge to glorify itself, it opens itself to destruction from within.

I was watching an old documentary—one of my favorites, *The Ascent of Man* by Jacob Bronowski—when an episode, "The Grain in the Stone," caught my attention. The whole series is about how humans developed this peculiar method called science. Bronowski doesn't follow the glib way of starting with Bacon and giving the usual superficial outline of hypothesis, experiment, and revision that you get in basic science textbooks; instead, he argues that science is central to the whole of human evolution, that what sets our lineage apart is an ongoing analytical relationship to nature.

For example, in this particular episode, the focus is on something that most science textbooks would not regard as part of their purview: the importance of art and architecture in reflecting a fundamentally scientific view of the world around us. The story begins with stone tools and ancient stone cities, which are more than just convenient stacks of natural objects; creating them actually required that the material be probed and manipulated and exploited. Bronowski suggests that this is the root of scientific thought.

> The notion of discovering an underlying order in matter is Man's basic concept for exploring nature. The architecture of things reveals a structure below the surface, a hidden grain which, when it's laid bare, makes it possible to take natural formations apart, and assemble them in new arrangements.

Dismantling things and reassembling them, testing ideas by implementing them, and furthering our understanding so that we can do something new—that's not just the world of the scientist; it's the world of the architect. Bronowski uses as his example the shift from the pillar-and-beam arch, to the round arch, to the high Gothic arch and flying buttress, so exuberantly exploited by the medieval church to build fantastic, open cathedrals; these changes required an understanding of the forces and the strengths in the stone. Even though architects in those times had no way to calculate forces, they had an intuitive understanding of how the materials worked.

A curious element of this program is that nowhere is it antireligious. At no time does Bronowski come out and castigate the follies of faith. Instead he does something far more insidious: he ignores it.

When he's discussing the triumph of twelfth-century Catholicism in providing money and coordinating regional communities to come together and build these huge edifices that strained the technologies of the day, he gives religion and the purpose of these cathedrals scarcely a mention.

The heroes are the craftsmen, the masons who used their knowledge and art to constantly push the frontiers of what could be done. This is only fitting. This is what lasts. The Parthenon was built as a temple to Athena, but Athena is nothing but a ghost, an empty symbol of a long-dead religion. What stands is a monument to the skill of the builder's hand and the art of the designer's plan. Catholicism isn't dead, but it is dying, just as surely as the Greek pantheon has faded, and the rotting nucleus of Christian faith is collapsing within a husk of its own construction—husks that are far more glorious and far better representative of human achievement than the noise and nonsense rattling about within them.

In financing those cathedrals, the church furthered its own decomposition by supporting an expansion of secular abilities. Where it thought it was shackling knowledge and skill in service to itself, it was actually funding applied research into properties of the material world, research that would lead in time to the Renaissance, the scientific revolution, and the industrial revolution. With hindsight, we can see that the church would have better maintained its stability if it had squashed human aspirations rather than encouraging them (and the church did play an important role as a central arbiter for the allocation of resources and talent in promoting the expansion of medieval thought). But we can't forget that religion is made by people for people, and people will value ideas that make their life better and easier—and religions will bend to adopt and use those ideas.

But then, aspiring is what we do. It's how we grow as a species, and it's antithetical to religious thought. Where religion thrives on stability and stasis, science thrives on the ragged edge of the known and unknown. Whenever we push back the darkness somewhere, we are killing a little piece of God.

But science is too often narrowly defined! It's not just people in lab coats with mysterious instruments—it's anyone who explores the world. And in this episode, Bronowski highlights artists, specifically sculptors, people who understand the hidden grain, and who take apart and rebuild structures to express an idea. The notion that artists are subversives and heretics is completely unsurprising, but great art is also about expressing something real through abstraction, representation, deconstruction, juxtaposition, revelation—a whole host of brilliant tactics for breaking through conventional perception.

Bronowski summarizes a key factor in what he calls the ascent of Man. It's not some amazing engineering triumph or complex scientific discovery—he's standing before a piece of abstract sculpture.

> The work can only be grasped by action not by contemplation. The hand is more important than the eye. We are not one of those contemplative civilizations of the Far East or Middle Ages that believe that the world has only to be seen and thought about, and who practiced no science. We are active, and indeed we know that in the evolution of Man it is the hand that drives the subsequent evolution of the brain. . . . The hand is the cutting edge of the mind. Civilization is not a collection of finished artifacts—it is the elaboration of processes. In the end, the march of Man is the refinement of the hand in action.

A dead civilization is one that has stopped progressing, that ends dynamism in the stasis of preservation and numbing reverence for the past. When a two-thousand-year-old myth becomes the greatest knowledge worth having, we have abandoned the process and begun to contract into the shells we built while we were still vital. We must continually break down and rebuild, reduce and reintegrate, disassemble and reassemble, in order to grow. I often hear complaints that damn "reductionist science" for breaking the world apart into smaller and smaller units—but those complaints are ill founded for two reasons. That destruction has always been and always will be a key part of human progress; contemplating what is, without breaking it to reveal what is within, is exactly the failure of the civilizations Bronowski is describing. And further, there is no such thing as purely reductionist science—science always builds again for deeper understanding. Evolutionary theory is a beautiful example of a powerful synthesis driven by insights from reductive analyses. It's an idea that is constantly being taken apart from within and rebuilt with new insights—its strength is in dynamism, not in stasis.

Blasphemers, heretics, and the sacrilegious are the cutting edge of the hand of humankind. Our eyes turn to everything and take it apart, picking at its bones and tracing the sinews and nerves to understand what makes it work, and the everything that we critically analyze includes the dogmas of our religious institutions. The most terrible thing we do to religion is to take it seriously—as seriously as we take a piece of flint, a stone arch, or a cathedral—and aim to take it apart, extract the bits that serve us well, and reassemble it into a tool that will serve humanity better. The chaff will be stripped, the nonsense will be carved out, the comfortable lies will be burned away, and what is left will not bear a trace of the revered superstitions, but will be a framework for human art and utility.

The Proper Reverence Due Those Who Have Gone Before

Some people might think I'm a rather morbid fellow. Years ago, when I was an undergraduate lackey at the University of Washington and working at the med school there, I made a wonderful discovery during one lunch hour: a bone room. Tucked away in an odd corner of the building was a room full of shelves stacked with cardboard boxes, each one containing the bones of some individual who'd left his or her remains to science. They'd been thoroughly cleaned and disarticulated, and many had parts sawed apart, so you could peer into the sinuses or the hollow spaces for marrow or poke around in the caverns of the cranium. It became my favorite quiet, private place. I could putter about reassembling someone, or just contemplate some scrap of bone for a Yorick moment.

Look at a humerus, for instance. It's elegant. You can see the traces of the muscle insertions that worked it in life. Its form is the product of a general genetic specification and a detailed, day-to-day remodeling in response to the forces that the individual applied to it. A pelvis and a vertebra are sculptures, intricate and odd. And a skull is a personal relic, a last vestige of a face someone knew well and loved without knowing all the wonderful knobs and seams and hollows buried under the flesh.

That's another thing; a bone isn't just beautiful operational engineering; it's a trace of a person. It's a melancholy memento of all that's been lost—here is this human being who struggled and loved and dreamed and hurt for sixty years, and all that I have of her now is a few exquisitely patterned swirls of hydroxyapatite. So much is gone, so much lost, and that's the fate of all of us. All it takes is a few generations for personal memory to fade away, and what's left is abstractions. Most of us won't even be bits of dry bone in a box in a forgotten room. We'll be ash and slime, our existence unremembered.

Maybe, though, while we are personally unacknowledged, some trace of us will be left in the genes of our several-times-great-grandchildren, or in a few words preserved in a library, or in some tiny nudge we've given history. That's all I aim for, to sow a seed that will in turn sow a seed that will sow a seed that—and so it goes. That's enough.

I am not a religious person by any means (that's an understatement), but I can feel a shadow of the same reverence for the Bible that I do for a piece of bone. It's a record, spotty and incomplete and flawed, of human lives, one that leaves out far more than it includes. It's not as pretty as a bone, but it's representative of some of the ugliness of human history, as well as some of the poetry. I can appreciate it as a slice of a few thousand years of the events and beliefs of one fairly influential tribe of people. A lot of lives and time, mostly unmentioned, are bound up in that book.

I want to try something, though, with the intent of getting across a point about the history of humanity. Let me reduce the Bible to an icon, a few pixels to stand for the whole thing, here:

|

Imagine that that is a Bible sitting on a shelf. My tiny black bar of pixels is a placeholder to represent everything in it, not to minimize it; if you have a grand view of the Bible's contents, that's fine; those few pixels should then conjure up your memory of historic events and aspirations and people who loved and raised families and created art and fought for what they believed in. And for those of us with less romantic visions of the Bible, it represents thousands of years of war and folly and pain and loss. No matter what, it's a big thing, a huge thing, and I've reduced it to a cartoon of the spine of a black-bound book for convenience. Just for now, keep in mind that it stands for two thousand years and the lives of hundreds of thousands to many millions of people.

Two thousand years, though, is a tiny slice of human history. In 1976 Mary Leakey discovered a remarkable fossil track in East Africa, the Laetoli footprints. Once upon a time, about three and a half million years ago, in present-day Tanzania, a volcanic eruption deposited a coat of fine-grained ash on the landscape, which was then wetted by rain to form a vast sheet, like firm cement, over everything in the region. Two, maybe three, people walked across the sheet, leaving their footprints behind in a material that then hardened in the sun, preserving their trail. We don't know anything about who they were, where they were coming from, or where they were going. We can imagine; they were walking together, one person larger than the other (a man and a woman? a woman and a child?), in a barren landscape wrecked by the volcano. Its eruption was certainly a life-changing tragedy, a catastrophe that upset everything they hoped for. They were living through a disaster of biblical proportions, and all we have left is a few lonely footprints. No other record of their life or their struggles remains.

These people were our very distant ancestors, small-brained and light-boned, but with a human posture. They were probably *Australopithecus afarensis,* and this earth-shaking event occurred 3.6 million years ago. It may be presumptuous to call them "people," "man," or "woman," since they aren't classified as human. But still, from what we know of our closest living relatives, the chimpanzees, they certainly cared and felt and thought, and australopithecines are more closely related to us than chimpanzees are, so we should recognize that they at least had something close to human feelings, complex social relationships, and personal wants and needs.

Another icon, even more famous than the Laetoli prints, is the skeleton of another australopithecine, Lucy. Just as we know little about the people whose relics ended up in those cardboard boxes from the bone room, we know little about Lucy the thinking, acting, living being. She was a small female, less than four feet tall, living in old Africa. We can imagine that she had family, she lived in a group or tribe, she foraged, she had hungry days and full days, she courted or was courted, she had moments of both happiness and grief. All the things she thought most important are gone and lost to knowledge, and all we have now are these few bones. When I hold the femur of a man who has been dead fifty years, I can feel the sorrow of a life lost to me. How much more reverence should we feel for these bones of a person from a world that has been gone for 3.2 million years?

And look at how much has been lost. Between the flight of the couple across a field of volcanic ash and the death of poor Lucy lies an interval of four hundred thousand years. If we define a Bible as a record of the struggle of a people for two thousand years, we'd need two hundred Bibles to tell us the tale of just this one obscure, remote branch of our lineage.

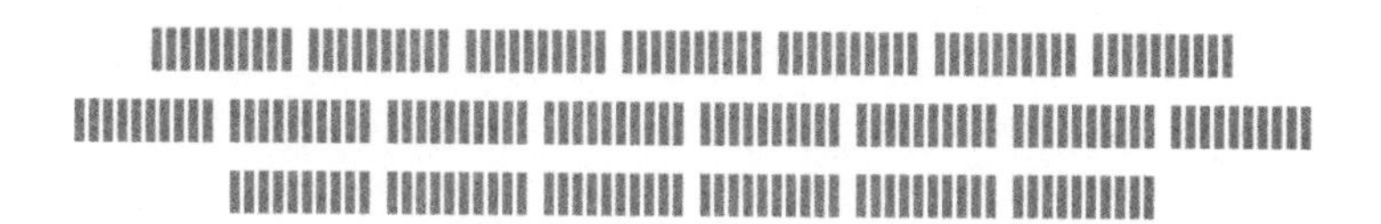

Two hundred Bibles that were never written, books that, even had they once existed, would be gone now. A vast history of events has been reduced to nothing but a few footprints and a scattering of bones.

Here's one more tragedy. (What's left to us is a record of the dead, and it's hard to avoid the sense that human history is one long tragedy. Joy is rarely preserved.) In 1984 a team led by Richard Leakey discovered another skeleton in Africa, this time of a young *Homo erectus,* whom they called Turkana Boy. We can refer to him as human, with no worry of picky quibbles that these are mere animal remains. These bones disturb my imagination. I'm a father; these are the bones of a young man, maybe twelve or thirteen years old, and he's tall and strong. In those dangerous days, the parents of such a robust boy perhaps felt relief that he was well past infancy, those risky years of high mortality; one would have been reluctant to become attached to an infant, likely to be carried away by some disease or brief famine. But Turkana Boy was a vigorous young fellow on the edge of adulthood, someone to carry on the line, someone to help on the hunt, someone to be proud of. And suddenly, he was dead.

You wonder—did his mother weep over him?

Turkana Boy died 1.6 million years ago. Between Lucy and this lost son, how many Eves weeping over dead Abels were there? Enough to fill eight hundred Bibles.

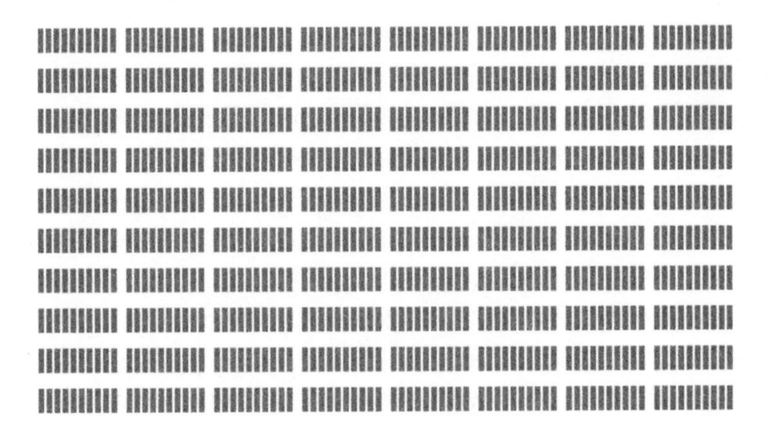

Think about that. Between the time of Lucy and Turkana Boy, our hominid ancestors evolved bigger brains, better tools, and more elaborate patterns of culture; they certainly fought battles, played out tragedies, explored new lands, raised children, and made romances, all of which are completely lost to us. This period of time was so great that our ancestors gradually changed physical form in response to the changing environment and their own mating choices—it was a genuinely epic story. I've mentioned just three known points in that story, fragmentary and mute as they are, but there is much more, museum cabinets full of broken bones and ancient tools.

Now here's the shocking thing; Turkana Boy takes us only halfway from Lucy to the modern day. To account for the rest, we would need eight hundred more books in this hypothetical lost library of humankind.

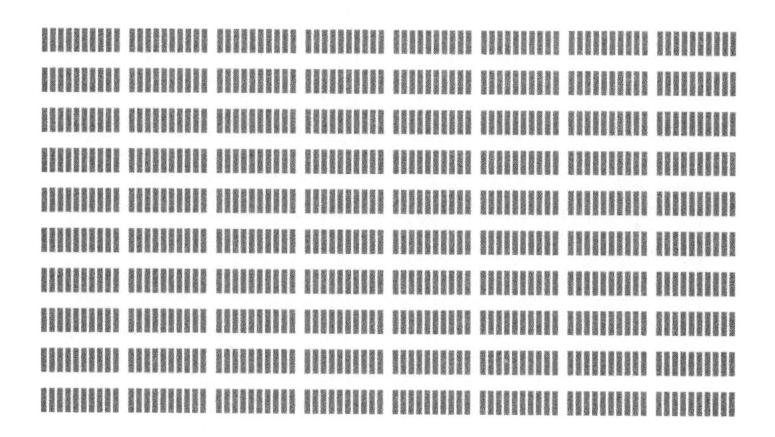

Remember, each black bar is an icon representing a long, elaborate book on the scale of the Bible, which in turn is only a small representative subset of the human experience over a span of time. So much has been lost to us, and those few scraps that we do have must stand in proxy for the grand sweep of history.

And you know, there are people now who claim that one book is sufficient, that it is complete, that it is enough to explain who we are and where we came from.

These holy books typically grant only a page or two, if that, to brief metaphorical sketches of the depth of human history. Christianity gives us only a pair of brief outlines that reduce everything before a historical line of Hebrew kings to an imaginary six days in the life of a creator, who brings everything into existence by fiat, without history, without precedent, without the mysterious pair in the dismal volcanic nightmare, without Lucy, without Turkana Boy, without any insight into the long, slow growth of humanity.

Strangely enough, these same people who bury our complex past under a veil of myth and denial claim to be "spiritual." To me, though, they are the ahistorical, unthinking ones who fail to offer the proper reverence due those who have gone before.

Niobrara

What do you think of when someone mentions the word *Kansas*? Maybe what leaps to mind is a farming state that is flat as a pancake; or if you've been following current events, the 2005 kangaroo court/monkey trial; or perhaps it is the drab counterpart to marvelous Oz. Kansas isn't exactly first on the list of glamorous places. I tend to read different books than most people, so I have a somewhat skewed perspective on Kansas: the first thing I think of is a magic word.

Niobrara.

Late in the nineteenth century, people stampeded to the American West to search for fossils of those spectacular beasts, the dinosaurs. Entrepreneurs everywhere were in on it—P. T. Barnum bought up old bones for his shows. Even scientists got caught up in the bone fever. Edward Drinker Cope of the Philadelphia Academy of Natural Sciences and Othniel Charles Marsh of Yale University were famous rivals in the bone wars, sending teams of men to Wyoming and Utah and Colorado and other Rocky Mountain states to collect the bones of the extinct terrestrial behemoths of the Mesozoic. Around the same time, Kansas was also a target, most famously by George and Charles Sternberg, for a different purpose: it was the place to go to find sea monsters.

There is a geological formation in Kansas called the Niobrara Chalk. Actually, it's not just in Kansas; it extends all the way up into Canada. But it has been exposed by erosion over much of northwestern Kansas, making it easy to dig into. And this is where the Sternbergs and Cope and Marsh went hunting for sea monsters.

Chalk is interesting stuff. It's made of a mineral, calcium carbonate, from the shells of microscopic, one-celled golden-brown algae. These Chrysophyceae are photosynthesizing organisms that float in large numbers at the surface of the sea, gather sunlight for energy, and scavenge calcium dissolved in the water to build their protective shells. They occasionally shed the minute calcium plates, and when they die, their skeletons drift slowly downward. A slow, soft, invisible rain of tiny flecks of calcium carbonate very, very slowly builds up on the sea bottom.

The Niobrara Chalk formation is six hundred feet thick.

It built up for a long, long time, tens of millions of years. The exposed chalks of northwestern Kansas date to between 87 million and 82 million years ago, near the end of the Mesozoic era and deep in the Late Cretaceous.

The inescapable conclusion is that Kansas was under water during the age of the dinosaurs. During the Mesozoic, the world was warm, and the oceans were at a high level, and central North America was a great, shallow, inland sea, a warm soup rich in microorganisms that were busily living and dying, and their remains slowly accumulated in deep dense chalk beds on the bottom. The North

American continent was split; the great inland sea ran up from Texas and into Canada and joined the Arctic Ocean.

It wasn't just coccolithophores living there, though. Shallow seas are fertile places for life, and this sea was home to vast shoals of fish and nautiloids, dense layers of bottom-dwelling molluscs and echinoderms, and amazing predators. Here lived the bulldog-jawed, snaggle-toothed fish Xiphactinus—over twenty feet long and eight hundred pounds of ferocious muscle.

Snaky-necked plesiosaurids feasted on the smaller fish. These genuinely weird animals—we have nothing comparable to them today—were diverse and successful, and their fossils have been found in large numbers in the Niobrara Chalk.

The predatory king of the Niobraran Sea was Tylosaurus, a mosasaurid that reached lengths of up to fifty feet. It was a giant, air-breathing reptile and was probably most comparable to a killer whale.

I've only briefly visited modern Kansas, but the Kansas of my imagination is a fiercely exotic ocean, a warm and savage sea richer than any place still extant. Try mentioning the magic word *Niobrara* to a paleontologist, or to any enthusiast familiar with Mesozoic reptiles—their eyes will light up as they conjure visions of the world of 85 million years ago, a world well documented in the incredible fossil beds of Kansas. That powerful, evocative word links us to a wealth of evidence and a complex, fascinating history.

In 2005 the Kansas State Board of Education ran a series of science hearings to give intelligent design creationism proponents an opportunity to make a case for the inclusion of creationism in the science curriculum. As I read about these ridiculous antievolution hearings, I found them sort of unbelievable. It isn't just that the creationist arguments were so poor, but that they were making them in Kansas, where beneath their very feet lie the relics of an ancient world that show them to be wrong. Didn't schoolchildren there take pride in the paleontological wealth of their home? I wondered. Did the people of Kansas bury their imaginations and avoid thinking about the history that surrounds them?

During the course of the hearings, the lawyer on the side of science, Pedro Irigonegaray, asked several of the witnesses for intelligent design creationism what they thought the age of the earth was. It's a simple, straightforward question with a simple answer: about 4.5 billion years. The intelligent design creationists found it difficult. Some of their answers were ludicrous, such as Daniel Ely and John Sanford's assertion that the earth is between ten thousand and one hundred thousand years old. Others were evasive: Stephen Meyer and Angus Menuge refused to answer. Some of these "qualified witnesses" were embarrassingly ignorant: William Harris could say only, "I don't know. I think it's probably really old." All of this is in line with the intellectually flaccid position of the godfather of the intelligent design movement, Phillip Johnson, who had bravely announced, "I have consistently said that I take no position on the age of the earth."

In Kansas there is a vast bed of chalk, six hundred feet deep, built up by the slow fall of dying microorganisms over vast millennia.

Mention the word *Niobrara* to these creationists, and their eyes will not light up. At best you might get dull-eyed incomprehension, and more likely you will see shifty-eyed evasion. Yet these characters want to dictate the scientific content of their children's educations. I swear, if there were any truth to their metaphysical codswallop, the shades of Cope and Marsh and the Sternbergs would have manifested in that courtroom to denounce them, and the floor would have cracked open beneath their feet to allow a spectral tylosaur to rise up and gulp them down.

The stones of Niobrara hold greater truths than do the dissembling and ill-educated brains of the fellows of the Discovery Institute, that intelligent design/creationism think tank in Seattle. We need to teach the evidence, not this phony, ginned-up controversy from a gang of poseurs and theocrats.

We Stand Awed at the Heights Our People Have Achieved

When the Buddhas of Bamiyan were dynamited in 2001, it wasn't an atheist who lit the fuse. We modern atheists who have stirred up so much resentment among the religious folk do not seek to demolish the past or to advance a destructive ideology. We aren't philistines who reject literature and art and music. We aren't monsters who would exterminate people to achieve our ends. And we aren't out to eradicate the world of ideas or to obliterate the vestiges of our religious history in art and architecture, although we have been accused of such nefarious plans. But such claims are easy to dismiss as the ravings of the delusional.

In 2007–9 Stanley Fish, the literary theorist and culture critic, repeatedly damned the New Atheists in a series of columns in the online edition of *The New York Times,* singling out notable atheists like Richard Dawkins and Christopher Hitchens. He didn't accuse us outright of conspiring to demolish Western civilization, perhaps to avoid the easily ridiculed paranoid martyr-complex of the mob. He didn't accuse us of being "new Communists," planning to march the orthodox to Siberia. Instead, he condemned us merely as unschooled near-illiterates with no appreciation of the depths of religious thought. We don't understand the nuances, he cried; we dismiss all the texts and traditions as "naive, simpleminded and ignorant." We just don't understand, period. He writes:

> Suppose, says Hitchens, you were a religious believer; you would then be persuaded that a benign and all-powerful creator supervises everything, and that "if you obey the rules and commandments that he has lovingly prescribed, you will qualify for an eternity of bliss and repose."
>
> I know of no religious framework that offers such a complacent picture of the life of faith, a life that is always presented as a minefield of the difficulties, obstacles and temptations that must be negotiated by a limited creature in his or her efforts to become aligned (and allied) with the Infinite. St Paul's lament can stand in for many: "The good that I would, I do not; but the evil which I would not, I do. . . . Who shall deliver me?" (Romans 7:19, 24). The anguish of this question and the incredibly nuanced and elegant writings of those who have tried to answer it are what the three atheists miss; and it is by missing so much that they are able to produce such a jolly debunking of a way of thinking they do not begin to understand.

Stanley Fish is a blind man.

First of all, that "complacent" faith that he claims doesn't exist is everywhere. Turn on your television. Watch a football game, and see the Christian players credit God with their touchdowns. Watch our politicians piously declare that they are praying for our troops. Watch televangelists milk their audience with blatant hucksterism; haven't you ever heard of the "prosperity gospel"? Give the good reverend a portion of your Social Security check and the wealth will come back to you tenfold. The bulk of religious thought is "naive, simpleminded and ignorant." Turn off your television and walk over to the local converted grocery store that is now being used by one of those fast-growing, evangelical/charismatic/Pentecostalist churches; ask anyone in attendance, and he will tell you (as long as you don't mention him by name) that Hitchens's characterization is correct. By faith will you be rewarded with paradise.

That second accusation, that we "do not begin to understand," is a subtler and more dishonest claim. Actually, it's clear that Fish doesn't understand.

What we atheists are saying is that we need to turn away from the powerless rationalizations of the holy books, no matter how poetic they might be, and recognize that their power and their appeal flows from their humanity, not from their religiosity. Forget God, that empty hulk, that great vacuum that humanity has stocked with its fears and dreams, and look at what we have created and felt instead. When someone weeps over a dead child or creates a great poem, it should matter not at all what some priest imagines his pantheon is doing. Take your eyes off your hallucination of heaven—what's real is that woman's tears, that child's triumph, that grain of sand, that bird on wing. Meaning is derived from the reality of what we see and feel, not from some convoluted vapor and self-serving puffery about the abstract concept *God.*

That baggage of superstitious, religious thinking is what

Richard Dawkins calls the delusion, and what Christopher Hitchens said "poisons everything"—not the book itself, not the literary qualities of the writing, not the pain expressed in the book of Job or the love shared in the book of Solomon. None of the prominent New Atheists wants the Bible burned or denied to readers. What we want is for people to think of it as a great hodgepodge of human expression that vindicates not so much a nonsensical image of a divine being as a complex, earthy, sometimes soaring, and sometimes hateful picture of us. But no, the true believers instead chatter inanely about the personal desires of that ludicrous god, the same kind of febrile blitherings that Fish offers as a justification for accusing atheists of not understanding the depths of the deity.

Let me give you an example of the godless view of your holy books. When I'm feeling the pain of grief, there's one book I turn to. I first found it when my grandmother died many years ago, and I've read it several times since, usually when I've lost someone I cared about. It isn't really a religious text, although it has gods in it, and a few religious concepts. But fortunately it's free of the mystical baggage with which our culture loads the Bible—it's easier to shake off the superstitious connotations when the god isn't named Jesus and isn't some being that your aunt believes is real, and there's no literary critic somewhere ready to accuse you of lacking an understanding of nuance because you don't believe in an old prophet's conjuring tricks.

The book is the epic poem *Gilgamesh*. I very much like the Herbert Mason verse narrative, which is less a literal translation (the original is a collection of 2,500-year-old Babylonian tablets) than an attempt to catch the spirit of the story. If you aren't familiar with it, put that Bible down and go read it—it's basically a lament about mortality. It's about a king, Gilgamesh, whose beloved friend Enkidu dies and leaves him wracked with grief.

> *Enkidu, whom I love so much,*
> *Who went through everything with me.*
> *He died—like any ordinary man.*
> *I have cried both day and night.*
> *I did not want to put him in a grave.*
> *He will rise, I know, one day.*
> *But then I saw that he was dead.*
> *His face collapsed within*
> *After several days,*
> *Like cobwebs I have touched*
> *With my finger.*

There is the touchstone, the common element that atheists and theists share, and that lies at the heart of those works that Fish praises and thinks the godless philistines neglect. We bond in the human experience; only the superficial and bigoted critic thinks some of us miss the point because we don't believe in the divinity of Anu and Marduk—or Jesus or Muhammad.

The rest of the story is a quest, as Gilgamesh tries to find a way to bring his friend back from the dead, or to win immortality for himself. There are deities and monsters involved, some interesting mythical figures, trips to the underworld, and so on—all quite thrilling stuff, as long as you're willing to recognize that it's all poetry and allegory and storytelling. What I, as an atheist, find most satisfying, though, is the reality. It's a story of loss. And most important, the quest fails—his friend Enkidu is not returned to him, and Gilgamesh realizes that he's going to die someday, too. He gets no glib, empty promise that if he only does the right rituals, then everyone will get together again in a paradisal afterlife. We live, we die, and people grieve; that's the hard truth.

> *In time he recognized this loss*
> *As the end of his journey*
> *And returned to Uruk.*
> *Perhaps, he feared,*
> *His people would not share*
> *The sorrow that he knew.*
> *He entered the city and asked a blind man*
> *If he had ever heard the name Enkidu,*
> *And the old man shrugged and shook his head,*
> *Then turned away,*
> *As if to say it is impossible*
> *To keep the names of friends*
> *Whom we have lost.*
> *Gilgamesh said nothing more*
> *To force his sorrow on another.*
> *He looked at the walls,*
> *Awed at the heights*
> *His people had achieved*
> *And for a moment—just a moment—*
> *All that lay behind him*
> *Passed from view.*

There's more to the godless view than just the "jolly debunking" and the assumed lack of understanding that apologists for faith attribute to us. We do not escape sorrow, we do not lack for joys, and we try as hard (I think even harder) as believers do to find meaning in our lives. We simply do not accept the shortcut of magical thinking that allows the lazy-minded to follow the path of religious escapism.

We stand awed at the heights our people have achieved. No gods, no religion. Us.

Acknowledgments

I have to give thanks to the professionals at the Knopf Doubleday Group who hammered the rough prose of a blogger with poor impulse control into some semblance of book-worthy coherence; to my editor, Jeff Alexander; to my agents, John and Max Brockman; and to my colleagues at the University of Minnesota Morris, who I think sometimes wonder who let the weirdo into the department.

For inspiration, I must express appreciation to those notorious atheists who blew open the doors for the public recognition of a self-respecting community of the godless: Richard Dawkins, Susan Jacoby, Daniel Dennett, David Silverman, and so many others. Loud and proud, people! For aggressive advocacy, I credit the crew at the Usenet site talk.origins, where I cut my teeth on the creation/evolution debate. For trying to keep me honest, thanks to the Horde, that diverse mob of vociferous commenters at my website . . . even if they do call me "poopyhead" all the time.

The deepest gratitude has to go to my parents, who in addition to providing a fortuitous arrangement of alleles in a pair of gametes most importantly brought me up in an open-minded family in which we were all encouraged to live freely and independently. It might be a little self-serving, but wow, they really did a good job parenting.

And, of course, I should share all the credit with my wife, Mary, who not only tolerates a hell-bound heathen husband but also worked late nights going over and over the text. She was completely indispensable.

Finally, no thanks at all to any god or spirit. Everything in this book was ultimately the product of the collaborative meat-brains of humanity.

A NOTE ABOUT THE AUTHOR

PZ Myers is a biologist, blogger, and sarcastic pain in the ass who lives in Morris, Minnesota. He is tolerated by his employer, the University of Minnesota Morris. (He guarantees you that they will roll their eyes in exasperation if you complain about him to them—they've heard it all.) He is regarded with benign amusement by his family, who have mostly adapted to his existence.